From Red to Green?

From Red to Green?

How the Financial Credit Crunch Could Bankrupt the Environment

Paul Donovan and Julie Hudson

publishing for a sustainable future

London • New York

First published 2011
by Earthscan
2 Park Square, Milton Park, Abingdon, Oxon OX14 4RN

Simultaneously published in the USA and Canada
by Earthscan
711 Third Avenue, New York, NY 10017

Earthscan is an imprint of the Taylor & Francis Group, an informa business

Earthscan publishes in association with the International Institute for Environment and Development

British Library Cataloguing in Publication Data
A catalogue record for this book is available from the British Library

Library of Congress Cataloging in Publication Data
Donovan, Paul, 1972-
From red to green? : how the financial credit crunch could bankrupt the environment / Paul Donovan and Julie Hudson.
p. cm.
Includes bibliographical references and index.
1. Environmental economics. 2. Financial crises. 3. Credit. I. Hudson, Julie, CFA. II. Title.
HC79.E5D655 2011
333.7—dc22

2011009431

ISBN: 978-1-84971-414-3 (hbk)

Typeset in Minion and Myriad by JS Typesetting Ltd, Porthcawl, Mid Glamorgan

Printed and bound in Great Britain by the MPG Books Group

For Andrew Hudson, Bethany Hudson, Charlotte Hudson, Emma Donovan, Lauren Hudson, Louise Donovan, Matthew Hudson, Samantha Donovan.

The next generation, who will have to pay for our generation's accumulated financial and environmental credit card bills.

Sorry.

Contents

List of Figures, Tables and Boxes

Figures

Tables

Boxes

Acknowledgements

It is hard to pin down exactly where the idea for a book on the twin credit crunches (financial and environmental) came from. The evolution from discussing environmental economics over coffee, to thinking about a book, to shaping the arguments and structure, has been lengthy. Along the way we have received a great deal of advice, encouragement and suggestions from a wide range of people. All of them have helped to push us towards this conclusion.

Larry Hatheway, UBS Chief Economist and Chief Strategist; Mark Steinert, UBS Head of Global Securities Research; and Nick Pink, UBS Head of European Equity Research, not only gave us their permission to write this work, but have offered considerable support for what has been (after all) a personal project. Mark Steinert and his regional teams lead a research department that actively encourages staff to follow their own intellectual pursuits and motivates its members to engage in critical thinking. Such a challenging intellectual environment is (sadly) relatively rare in financial markets today, and we both recognize our great good fortune in working for an organization that realizes the benefits of rigorous research.

George Magnus, UBS Senior Economic Adviser, gave a great deal of his time and intellectual and publishing experience in getting this project off the ground. He motivated us to write a book in the first place, and was generous in offering support, even while penning an economics book of his own (looking at the consequences of the financial credit crunch for emerging markets). We hope that *Uprising* will prove to be the *second* best selling credit crunch book of 2011.

Erika Karp, Head of Global Sector Research at UBS, directed some of her seemingly limitless enthusiasm our way. Her faith in the people who work for her and what they can achieve was a powerful antidote to the occasional despondency that can afflict even economists (and never seems to be far from environmental research).

Sir David King, Chief Scientific Adviser to UBS and Director of the Smith School, Oxford, very kindly offered both comments and encouragement for this project, and also read through an early draft of the opening chapters with great tolerance and without complaint.

Emma Howard Boyd – Head of Socially Responsible Investment and Governance, Director, Jupiter Asset Management Limited, and an independent non-executive member of the Environment Agency Board – willingly took a draft of part of the book with her on her travels, and provided us with a very helpful preliminary review. Patrick O'Bryan willingly sacrificed time and energy in giving the book a trial run. Ruth Ridout and Dorothea Hill also gave us some valuable no-nonsense perspective.

In looking at so wide-ranging a topic as the consumer, we have had to lean heavily on the expertise and assistance of others. Alex Robson came up with a great deal of valuable research on the environmental impact of consumer electronics and whose grasp of other fast hard facts was remarkable. Ilsa Colson was extremely helpful in giving information and sources for the chapter on water. The time she spent working with Senator Penny Wong (then Minister for Climate Change and Water in Australia) gave her a great deal of knowledge (matched only by her detailed knowledge of Melbourne restaurants). Chris and Judith Trimming imparted a great deal of practical information on rural economics and wood-burning stoves.

Countless other colleagues, clients and friends have been generous with their views – either in lengthy discussions, or snatched conversations. We have benefited enormously from all of them, but special thanks go to Scott Haslem, Jeff Palma, Rachel Baird, Ruth Girardet, Shirley Morgan-Knott, Hubert Jeaneau, Gbola Amusa, Joe Dewhurst, Peter Hickson, Stephen Oldfield, Jon Anderson, Alan Erskine and many others; to Earthwatch scientists Dr Pete Kershaw, Dr Mark Huxham and Professor Weizu Gu; to the Barichello family of Dechenla; to all of the board members of the Research Foundation of CFA Institute past and present; and to several Warwick University English professors (MA programme 2010–2011), who have (knowingly or unknowingly) helped form the ideas in this book.

Finally, we should offer our thanks to the fellows and staff of St Anne's College, Oxford. Quite coincidentally, we both attended St Anne's (at different times), and the culture of the college has undoubtedly shaped the way we view the world. The royalties from this book are being used to help fund environmentally friendly aspects of the college's building programme.

Of course, any errors or omissions remain our responsibility. The great advantage of being co-authors, however, is that we both have someone else to blame if any errors have crept in.

PREFACE

From Red to Green? A Tale of Two Credit Crunches

Anyone who believes exponential growth can go on forever in a finite world is either a madman or an economist. (Widely attributed to Kenneth Boulding, an economist)

Introduction: Consumers and the environment in more constrained times

Virtually the first thing any economics student learns is that economics, ultimately, is concerned with allocating scarce resources among infinite desires. One of the first things any student of the environment will come to realize is that, in making this resource allocation, all economic activity will have some kind of environmental impact. Consuming resources, or choosing not to consume resources, has a bearing on the environment.

We believe that the world economy is going through a significant structural change in a strictly economic sense, as it comes to terms with the consequences of the financial credit crunch. Economic norms are being changed, in a manner that the world (at least the developed economies) has not experienced for a generation.

At the same time, issues of environmental sustainability are becoming increasingly important, in terms of both resource constraints and the lasting environmental consequences of using those resources. Effectively, there is a second credit crunch – a nascent environmental credit crunch – which has virtually identical structural characteristics to some key aspects of the financial credit crunch. These two simultaneous crises have a symbiotic relationship, which we believe will lead to a significant adjustment in the way that consumers develop in the future. This book aims to examine how consumption is likely to change in the wake of the financial credit crunch, and how that change will impact environmental sustainability (the availability of environmental credit, in effect). At the heart of this process is the way in which environmental and financial credit works.

Credit

So what is credit? Credit is simply the ability to consume future resources today – at the expense of consuming in the future. A mortgage allows a borrower to buy a house without a huge financial outlay, and then to consume (occupy) that house. In exchange, the borrower surrenders some of their future income (which means their standard of living) each year, in order to repay the credit. The result is today's standard of living rises, assuming that owning a home raises one's standard of living, and tomorrow's standard of living is lower than would otherwise be the case, because of the future income foregone.

Conventionally, credit is thought of in financial terms, as economic transactions. However, it does not have to be. It is also appropriate to think of environmental credit. Any finite resource (a barrel of oil, for instance) that is consumed today cannot be consumed in the future. The consumer has taken a decision to raise their standard of living today at the expense of enjoying a higher standard of living tomorrow.

The environment is never quite as clear as the pure world of economics, and there is a subtle variation for environmental credit. Human actions can impact (generally reduce) the future availability of environmental resources by changing the way in which the environment works. Climate change would be one example, or a species becoming extinct. That process will also reduce the ability to consume environmental resources in the future, as a consequence of today's consumption decision. Hence, should a fish species such as cod become extinct through overfishing today, future consumption of that much-loved culinary delicacy, cod and chips, would be zero tomorrow. The financial parallel to this form of environmental credit is, perhaps, spending so recklessly on a credit card today that no sane bank will lend you money in the future (as your credit rating is slashed, destroying your ability to borrow).

The concept of environmental credit can be summed up in the idea of the Earth Overshoot Day, devised by the New Economics Foundation. This date is calculated each year by the Global Footprint Network. To quote the Network's website, Earth Overshoot Day is the day when 'human demand on ecological services begins to exceed renewable supply'. The following chart shows the number of days the Earth is in 'overshoot' (i.e. how many days in each year that the Earth is unsustainably consuming resources according to this perspective on the ecosystem).

In 2010, Earth Overshoot Day fell on 21 August. On this day, the Earth's production of renewable resources for the year had been used up by current consumption patterns, leaving human beings living on (environmentally) borrowed time for the remaining 132 days of the year.

This environmental borrowing is, ultimately, more unsustainable than the consumer borrowing that preceded the financial credit crunch – this is a super

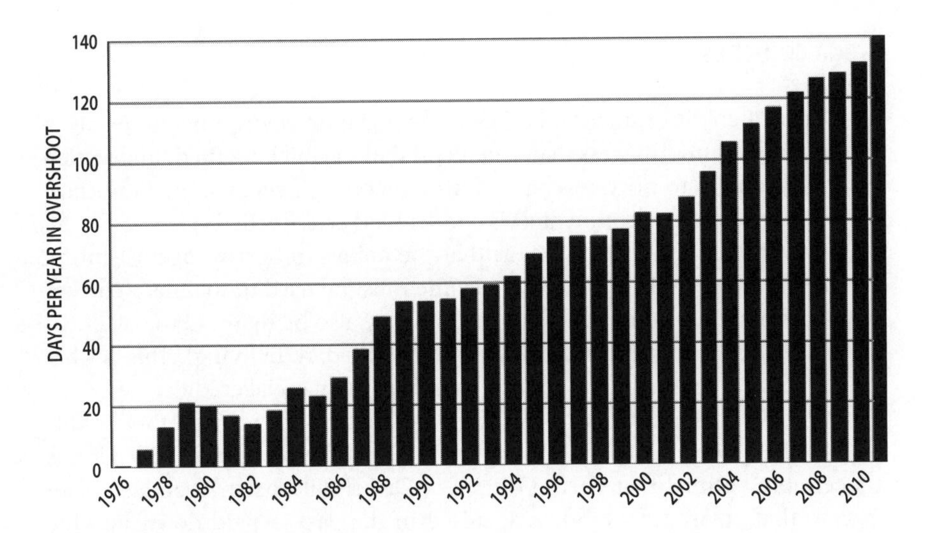

FIGURE P.1 Earth Overshoot Day

Source: www.footprintnetwork.org.

sub-prime environmental credit problem. For a third of the year, in 2010, the human race was consuming tomorrow's resources today, with a vague promise to settle up the bill in a few years' time. Moreover, the problem appears to be getting worse: there are predictions that by 2030 we may well need two planets to sustain us (or half the year will be spent living on environmental credit).[1] The idea that someone could live entirely on credit card borrowing for half a year would strike most people as an unsustainable situation. If it is unsustainable in terms of financial credit, it must be even more unsustainable in environmental credit terms, for, unlike finance, the environment is unlikely to offer technical get-outs such as the financial restructuring known as chapter 11 bankruptcy in the US. This point drives home the message that, if we do not slow down our consumption of 'renewable' resources to a more sustainable pace, life in at least some regions of our planet may become uncomfortable at best, untenable at worst.

A credit problem never seems like a problem while the credit continues to flow. The fatal phrase 'this time it's different' is brought out to justify why credit can continue indefinitely, and why the party need not stop. But, at some point, the party has to stop. At some point, credit is crunched. That point, financially, occurred at the end of 2008. That point, environmentally, lies just a short way in the future. Indeed, the environmental credit crunch may already have started.

Credit crunches

So what is a credit crunch? Whether in financial or environmental terms, a credit crunch is nothing less than a demand that the bill be settled (and settled now). The ability to borrow against future income (if financial) or to utilize finite resources (if environmental) is constrained or crunched.

In economic terms, this means that the ability to borrow against future income is reduced. More of today's income must be used to maintain current standards of living. Some of today's income may also be required to pay down yesterday's debts (finally to pay for yesterday's standard of living). This is what the world is experiencing today, when it comes to financial credit.

In environmental terms, a complete credit crunch (i.e. no more credit) requires Earth Overshoot Day to fall on 31 December each year. Indeed, it is conceivably optimal that Earth Overshoot Day should occur even later than that, so that environmental debt acquired in the past is 'paid down' and the environmental damage of previous years is repaired.

A tale of two credit crunches

The timing of the financial credit crunch has coincided with a greater understanding about the dangers of the environmental credit crunch. Consumers, in developed economies at least, are learning that they cannot live beyond their financial means indefinitely. While this financial lesson in prudence is being learned, however, we still find that consumers are living beyond their environmental means. This cannot go on indefinitely, and the interaction of these two credit crunches is critical to how the world develops – in an economic and an environmental sense.

Organisation for Economic Co-operation and Development (OECD) economies have seen a significant increase in consumer spending over the past two decades. The increase in financial credit availability from 1990 onwards removed constraints on consumers. Globalization (which nearly doubled in scope over the past two decades) broadened the range of goods available to all income classes. Asset values rose as risk premia declined, which added to the affluence of specific groups in society. Consumers across the world are increasingly aware of the range of goods that they can aspire to owning, as information improves (a shift that can be compared with earlier consumer revolutions, such as late 18th century Britain).

For some goods, environmental sustainability has become a luxury that consumers aspire to (such as organic food), especially if the associated cost difference is not too great. At the same time, the fact that environmental costs (where they are included in the price of the product) form a small part of overall consumer spending may mean that consumers may feel happier about ignoring the environmental consequences of their actions. Consumers might feel 'happy' to throw away a plastic bag that they have paid five pence to purchase because that five pence levy has salved their environmental conscience. The consumer does not need to feel guilty (they have paid the price), but because the price is so small the consumer has little incentive to reuse the bag (as well as feeling no guilt about throwing it away).

The world is now changing. Economically, we believe that consumer credit will be less readily available. Global economic growth is likely to have a lower trend rate. Globalization is likely to stagnate. Consumers are therefore going to have to moderate their rate of consumption increase. At the same time, both through economic effects (for example, the pricing mechanism, or income effects) and visible environmental consequences (such as extremes of weather), we expect consumers to be increasingly aware of the environmental credit issues. We have reached a critical point. The financial and environmental bills are arriving at once.

So what do the twin credit crunches mean?

This book is an attempt to examine how the 'red' overdue bills and final demands of the financial credit crunch will impact the 'green' of the environment. Taking different sectors of consumption, chapter by chapter, we examine how changing financial and environmental credit conditions are likely to alter the pattern of consumption in the future. In some instances, the consumer behaviour changes arising from the financial credit crisis are likely to promote environmental sustainability. In some instances, environmental sustainability could be harmed by the changes, in the absence of government action. Our conclusion summarizes both perspectives.

Finally, we draw some conclusions for broad trends in the world economy, and how policymakers might act to maximize both economic growth and environmental sustainability in a changing economic and environmental climate.

What is certain is that the consequences of the combined effects of the twin credit crunches cannot be ignored. If overdue bills are ignored for too long, the bailiffs are called in to seize assets and exact payment.

No one wants that.

Note

1 WWF in collaboration with the Zoological Society of London and the Global Footprint Network (2010) *WWF Living Planet Report*, p9.

List of Abbreviations

ABS	alkylbenzene sulfonate
BOGOF	buy one get one free
BRE	Building Research Establishment
CHP	combined heat and power
CPI	consumer price index
CRC	Carbon Reduction Commitment
CSP	concentrated solar power
Defra	Department for Environment, Food and Rural Affairs
EPA	Environmental Protection Agency
EPC	Energy Performance Certificate
EU	European Union
FAO	Food and Agriculture Organization
FMCGs	fast-moving consumer goods
FSC	Forest Stewardship Council
GDP	gross domestic product
GHG	greenhouse gas
HVDC	High Voltage Direct Current
LAS	linear alkyl sulfonate
LCA	lifecycle assessment
NGO	non-governmental organization
NHS	National Health Service
OECD	Organisation for Economic Co-operation and Development
OPA	Office of Price Administration
R&D	research and development
REACH	Registration, Evaluation, Authorisation and Restriction of Chemical Substances
UN	United Nations
UNCED	United Nations Conference on Environment and Development
UNDP	United Nations Development Programme
UNEP	United Nations Environment Programme
VOC	volatile organic compound
WHO	World Health Organization

CHAPTER 1

Food in the Face of the Two Credit Crunches

I by no means take those gloomy views of our prospects which are entertained by a large number of my countrymen. Such views, I admit, are perfectly justifiable if any attempt is made to plod on with a system which was very suitable to the conditions of twenty years ago. (R. H. Elliot, *Agricultural Changes*, 1898)[1]

At the risk of asking an obvious question, why is food so important? Why start a book on consumer spending with the topic of food? For developed economies at least, this must be about more than the bare necessities of life – most wealthy consumers passed the point where calorie intake constitutes one of the bare necessities of life several pizzas ago. In spite of this, food clearly is important. Food prices make headlines in the press. The topic of food agitates politicians. Even in societies where most are free from want, issues regarding food have led to riots in recent years (for example the February 2007 protests over tortilla prices in Mexico). Why is the topic of food so important to consumers – and what are the implications of the collision of the financial and environmental credit crunches on so critical an area?

The importance of food

Politicians in all but the most despotic regimes put a premium on the provision of affordable food for all: food, economics and politics have always been closely associated. The British political class of the mid-19th century was intimately connected with the land (often as farming landowners), and had to be. A bad harvest foretold political tension and social unrest because it would herald food shortages with significant increases in the price of staple foodstuffs. The British general election of 1906 was known as the little loaf election – and was won by the Liberal Party, whose claim that the imperial preference trade policy of the Conservatives would lead to a 'little loaf' of bread for consumers (in contrast to their own 'large loaf', made affordable by free trade).

The fear of food shortages and food expense still resonated with the electorate years after the Industrial Revolution. Even in the era following the Second World War, food pricing has preyed on politicians' minds. US President

Nixon was ideologically opposed to price controls (scarred by his experience of administering them in President Roosevelt's Office of Price Administration (OPA) during the Second World War. Nevertheless, he embraced them in 1972 because of an obsessive concern about the American housewife's perception of the cost of her basket of goods.[2] Indeed, it has been suggested that food price inflation was 'the issue that did more than any other to destroy Nixon's reputation for competent economic management' (Bowles, 2005, p132).

This historical spectrum nicely demonstrates the political problem when it comes to food. At one extreme is the issue of a real food shortage accompanied by hunger. The other extreme echoes to the cry of middle-class outrage – consumers who see food price inflation as 'disgraceful' but are not driven to starvation as a consequence. Such is the nature of democracy that the consequences of domestic middle-class outrage easily (and frequently) triumph over foreign hunger in the minds of politicians. Politically, food matters.

The economics of food pricing

Why do food prices have so powerful a role, even in wealthy societies? Of course, food is a basic support for life, and want will provoke political hostility. On the other hand, food is a relatively small part of the family budget nowadays (food and non-alcoholic drink spending for home consumption is less than 11 per cent of the UK consumer price index (CPI) basket of goods) and, in a developed economy, real hunger is no longer a credible concern for most.

The issue is that spending on food has a wholly disproportionate influence on the way consumers see the world. For all that economists like to pretend that we live in a perfect world of rational consumers, the sad fact is that reality does not live up to the high standards that economists set. Consumers are not rational – not even close (economists often feel that they alone are the last redoubt of rational sense in an increasingly irrational world). *Irrationally*, the things that consumers purchase on a regular basis shape their economic perceptions out of all proportion to their actual importance in the household budget. Consumers are sensitive to the price of something that they purchase daily, for example, because every day they are reminded of the price. Something purchased once every two years, though it may be a larger part of their overall spending, is 'out of sight, out of mind'.

There are few things that consumers purchase more frequently than food. A change in the price of an item bought regularly, or a rise in the overall cost of a weekly shop (whose basket composition remains relatively static), will be noticed and provoke the middle-class outrage that politicians so fear. The price of a chocolate bar rising from 40 pence to 50 pence in a workplace vending machine will lead to fulminations of inflation being out of control. Of course, chocolate is hardly central to a consumer's basket of purchases and, relative to income, the price increase is negligible. However, the chocolate bar is a regular purchase. Mid-morning, every working day, the consumer is reminded that prices are rising. The perception of general price increases is thus established and becomes the prism through which the wider economy is viewed. Surveys of consumer perceptions of inflation show a strong correlation not with the actual calculated consumer price inflation number but with the food and energy parts of the price index.

Why food pricing fuels the environmental credit crunch

This sensitivity to food prices works (through the political process) to impact the environment. Over the past half century, the short-term election cycle (focusing politicians on food prices now rather than food prices in decades to come) together with globalization (which allowed market forces to deliver food price deflation to customers) have combined to change the way food is procured, transported, processed, merchandized, prepared and consumed for significant swathes of the global population, mostly in developed countries.

This drive towards low-to-negative food inflation for large numbers of relatively well-off people has had unintended consequences, in the form of increasing environmental and social costs. Consider the increase in food waste, for instance. Contemplate the rising obesity prevalent in wealthy countries. That the system has conspired to throw away food, have an obesity epidemic and experience hunger in parts of the world *at the same time* is truly astounding (clearly, not astounding in a 'good' way). As food markets are now far more global than they were historically speaking, mere food price sensitivity in one country may well raise the prospect of starvation in another, causing politicians to act in potentially irrational ways.

Politicians' search for negative food price inflation (and the glittering prospect of electoral success that might trail in its wake) has, more than anything else, led to the generally unrecognized consequence of 'environmental cost inflation'. The balance between controlling short-run financial costs and longer-term environmental costs has been firmly skewed in favour of the short term. In effect, politicians are happy to have society run up an enormous bill on its environmental credit card, if it means short-term economic benefits (and the associated political benefits). Thus, the environmental credit crunch looms ever larger, because of the short-term economic focus.

The environmental challenge ahead (and it is huge)

One of the greatest challenges humanity is likely to face in light of the United Nation's (UN) world population projection of 9 billion by 2050 is the adequate provision of food. Land on which food can be grown or animals farmed, and water bodies in which fish can be cultivated or harvested, are limited. For all that has been done with technology in agriculture, experience to date suggests that the productivity of natural resources, such as soil, plants, and fish stocks, also has limits.

As the human population grows and puts increasing strain on the land, water, air and organic matter that support food production, we can expect this to appear, somewhere, in market prices. Thus, the short-term desire to achieve low food prices hits the global environmental credit crunch, and ultimately becomes self-defeating: keeping food prices low today adversely affects the environment, which (in turn) will potentially constrain our ability to produce food in the future, thereby raising food prices in the future. Global food markets in the context of global resource constraints will inevitably produce social injustice somewhere.

A number of writers refer to a particularly important constraint facing the agricultural sector: this is the tendency of the human population to grow exponentially, while the productive capacity of the agricultural sector is only capable of linear growth. In the presence of a growing population, the globalization of agricultural commodity markets can (and has) helped to fill the mathematically driven supply–demand gap from one region to the next. Population controls driven by public policy (for example, as implemented in China, Japan and Italy) may well be an alternative approach to the problem. Balancing supply and demand and managing food security risk are very old problems. What is new is the sheer scale of the human footprint on planet Earth, and herein lies the challenge.

The question is how global politics will deal with this. The climate change talks (Kyoto, Copenhagen, and everything in between), that have increasingly gained in profile since the early 1990s, may address some of the environmental issues relevant to the apparently irrational behaviour of national political cycles that disregard the longer term. The frameworks that come out of the talks may also deliver a political paradigm that is at least partially fit to deal with the unique challenges currently facing the food and agriculture sector.

The changing economic environment for the consumer that is emerging from the wreckage of the financial credit crunch plays directly into the importance of food spending. This is certainly true from the perspective of food prices in the short run, but possibly also from the perspective of sustainability and reducing the impact of the environmental credit crunch. The economic changes we experience now have a bearing on our ability to manage the environmental credit crunch that looms. The economic changes we experience now may also have a bearing on the political willingness to take the hard decisions that will be necessary to minimize the long-term economic and environmental consequence of the environmental credit crunch.

This is a mess. The financial credit crunch is bad enough on its own but it is taking place at quite the worst possible time in the context of the environmental credit crunch. The rest of this chapter looks at how consumption of food is likely to change as a result of the financial credit crunch, and what the implications of those changes are for the food environment. There are four key areas of change: efficiency, substitution, waste and food miles. We then consider whether these consumption changes are likely to be enough to reverse the causes of the environmental credit crunch (as we saw in the chart in Figure P.1, page xv) in food, and, if not, what else needs to be done.

More calories for the buck

The economics

The financial credit crunch is reducing the standard of living previously enjoyed by the consumer. Every time their credit card is declined at a checkout, the consumer loses a little of a lifestyle they had previously assumed was secure. Confronted with this extreme, the wily consumer will seek to find a way to spend less money without getting substantially less benefit. For food that means getting the same calorie intake for less financial outlay; one way would be sacrificing more expensive protein for bulky carbohydrates. The famous English dish, roast beef and Yorkshire pudding, was served pudding first (to fill everyone up before the meat was served) for exactly this reason. Mr Holbrook, in Elizabeth Gaskell's *Cranford* (published in 1851) recounts the dining 'order of precedence' of his youth: 'No broth [soup], no ball [suet pudding]; no ball, no beef'. Carbohydrates came first in a strict hierarchy; the Atkins' diet was an anathema.

With obesity a very public issue in the West, getting sufficient calories may seem to be a remote problem. Hunger is, however, very much a problem in developing countries, and is often accompanied by micronutrient deficiency. Some of the indirect effects of the financial credit crunch may be a rise in poverty that is sufficient to reduce the food spending for some.

By reducing the amount spent on food consumption, the consumer will end up with more money to spend on other items, or to repay debt. How can one reduce the money spent per calorie consumed? In fact, the change can be effected with surprising ease, and is undertaken almost instinctively by consumers in Organisation for Economic Co-operation and Development (OECD) economies when they are confronted by an economic downturn. Developing economies' consumers are more directly impacted by changes in the price of agricultural commodities. Developed economies' consumers can change their consumption more readily because, in a developed economy, food is hardly food at all.

Let us consider a takeaway hamburger meal from a fast-food chain. A consumer might spend (for the sake of argument) £3 on an upgraded cheeseburger, French fries and soft drink. If feeling particularly reckless, they may even enlarge the meal for an extra £0.50 – but let us assume restraint (there is a financial credit crunch to consider, after all). What is it that the consumer is purchasing with that £3? What the consumer is not buying is £3-worth of food. In fact, the food content of the money spent is pretty minimal. What the consumer is buying is perhaps £0.60 of food produced by a farmer. The remainder of the food bill is paying for everything that happens from the point at which the agricultural produce leaves the farm gate, and ends up in the fast-food outlet. This includes the processing of the raw product into processed food; the manufacture of the packaging (including the paper pulp that goes into the packaging); the advertising and marketing of the fast-food chain; the cost of transporting all of this to the restaurant (including the cost of the driver, the cost of purchasing or leasing a lorry and the fuel costs involved); the cost of the fast-food chain's management; the cost of the kitchen staff; even the cost of the investment bankers who advise the fast-food chain will end up reflected in the £3 that is handed over in exchange for the cheeseburger meal. In other words, the consumer is paying not for food, but for the extraordinary amount of labour involved in processing food (plus a few other raw materials along the way).

When, in 2008, food prices started to rise globally, some of the more alarmist commentators sought to put the blame on China. Here was a convenient scapegoat – a country of more than a billion people, whose spending on food was rising at a rate of between 20 per cent and 30 per cent a year. China was eating the rest of the world into a global food shortage. This was, of course, nonsense. Chinese food consumption, measured by calories consumed, has been rising at a very modest 2 per cent or 3 per cent per year in the recent past. What China was doing was upgrading to cheeseburger meals (metaphorically and, in some cases, literally). Spending on food was rising because China was spending more on the preparation, packaging and presentation of food. The actual food calorie consumption was changing very little.

This then gives a pretty broad hint as to how consumers can maintain the benefit of food while cutting spending. The first element of trading down is to reduce the labour content of food spending. Rather than going out for a meal at a pub, the consumer buys a steak from a supermarket and cooks it at home as a response to the financial credit crisis. They no longer get the service of having someone else cook the steak for them – but then they no longer have to pay for that service. The money saved assumes disproportionate importance in the eyes of the consumer, because a relatively high-frequency purchase has been altered. The consumer has the virtuous sense of having made a significant economy in their household budget, but does not feel that they have compromised their standard of living too much in the process.

In fact, the aftermath of the financial credit crisis has seen exactly this shift in consumption patterns coming through. In 2008, spending on food and drink consumed at home dropped 0.4 per cent in the UK (after accounting for inflation); it fell 0.9 per cent in 2009. Spending on food consumed in restaurants fell 3.4 per cent and then a further 6.5 per cent in those two years.[3] A consistent pattern of economising in food is evident as the financial credit crunch took hold.

This change in consumer behaviour has environmental consequences, of course. In particular it impacts energy demand. In the US, analysis of the food supply chain suggests that home preparation of food (in the average household) accounts for 25 per cent of the total fossil fuel consumption associated with food production and distribution. However, food prepared at home is more energy efficient than food prepared at a restaurant. US consumers receive somewhere between 8 per cent and 14 per cent of their calories from restaurants, but the restaurant sector consumes 39 per cent of all food preparation related fossil fuel consumption.[4] Restaurants are not just a drain on consumers' credit cards, they serve as a significant drain on the environmental credit that society has been drawing down.

Food consumption patterns are also shifting at home. If a consumer is seeking to trade down, moving from restaurant spending to home cooking will not be the only economy that is sought. Home food spending patterns will also adjust. If we look at a raw material level, meat is the most expensive part of a typical consumer's food basket. Meat spending accounts for 23.5 per cent of the average UK consumer's overall spend on food consumed at home. This is appropriate as it reflects the input costs associated with meat. The expense of meat means, however, that in less affluent times, meat has played a smaller role than it does today. In the period from 1961 to 2002, meat consumption per head nearly doubled globally. As both developed economies and emerging markets became more affluent, meat consumption rose.

The reverse can also be considered. That which we consider a basic foodstuff today can quickly assume the role of a luxury tomorrow. In mid-18th century Scotland, 'the meanest servants not at board wages will not make a meal upon salmon if they can get anything else to eat' (Russel, 1864, p92). Salmon was then sold for a penny a pound. By 1805, salmon was 'generally too dear to be used by the common people' and the Duke of Richmond received £5000 per year for his Spey salmon fishery rights. The commonplace food of history became a luxury within a generation. In the 20th century, the pendulum swung to and fro again. Fish farming meant that salmon became plentiful, and the abundance of supply meant that its price fell. Then, the intensive nature of fish farming led to an increase in disease in the farmed fish stock, which reduced supply, and the price of salmon began to rise once more.

Meat as a staple of a person's diet is a very recent concept. An inmate of a London workhouse in the mid-19th century would have been amazed at the idea of being proffered meat as a daily entitlement. Charles Dicken's beadle, Mr Bumble, reacts with horror to the idea that Oliver Twist has been fed meat by his employers, the Sowerberrys – and this a work of fiction that is well grounded in reality. A workhouse inmate had 'meat days', but these were far from being every day, and their calorie intake was far less meat dependent than in today's society.

The role of meat in a consumer's diet does not have to alter that radically to have a profound impact on the environment – because the impact of

TABLE 1.1 Weekly calories for various diets versus a serving of double cheeseburger and fries today

Food source	*Meat*	*Sugar*	*Vegetables*	*Bread (UK 1833/1945) or cereal and pulses (world)*
1833 British workhouse pauper	1023		2968	6320
1833 British soldier	3580			2374
Second World War British rations	1002	467		
World average 1961–1963	483	644	1932	10,626
World average 2001–2003	1372	1176	2156	10,584
Single serving of quarter pound double cheeseburger, large fries, soft drink	623 including cheese		522	120

Source: Authors' calculations[5]

meat on the environment is so particularly intrusive. Turning back the clock by even a generation would have positive environmental consequences, if the economic situation leads to a reduction in meat consumption.

The environment

Meat is an expensive source of protein from all perspectives. Even with the relatively high economic price of meat, the environmental resources required to produce it are not necessarily fully reflected in that price. With modern factory farming techniques, it takes 33 calories of inputs to produce one calorie of beef. Of course, the *market* cost of those calories is reflected in the price of beef on the supermarket shelf (otherwise the farmer would be working for a loss). The consumer of beef is paying for each of the 33 calories as they consume their one calorie of rare topside – which is why beef is so (economically) expensive. What is missing from the price is the full *environmental* cost of the sort of energy that goes into rearing modern-day beef.

Agriculture has met the increased demand for meat with an increase in supply. The problem is that the grass-nibbling, oat-munching, snorting, muck-spreading plough horse has been replaced by a fuel-hungry, engine-noise-emitting, fume-snorting tractor. Not only does the tractor pollute the air as it works, its use removes the recycled waste product of the horse – the manure – and requires the substitution of chemical fertilizers. This technological progress therefore forces an increase in the consumption of *non-renewable* resources – drawing down heavily on environmental credit. A plough horse's fuel of choice (grass) is a readily renewable energy source (it is converted solar energy, in effect). A tractor's fuel of choice (diesel), once consumed, is lost to the global economy. This is the environmental credit crunch at its most obvious. What has the plough horse and tractor got to do with the production of meat? The United Nations (UN) has estimated that 60 per cent of maize and barley production between 1961 and 2001 went to animal feed. The farmer may plough the fields and scatter, but the good seed generated is generally converted into meat.

The agricultural process is fundamentally changed by its increasing dependency on finite resources. Our food supply no longer depends on the flow of energy coming from the sun, but also from the stock of energy stored in the earth (in the form of fossil fuel and other deposits). Food, as currently farmed, is a finite resource, not a renewable resource. When this is put in the context of the environmental credit crunch, the global challenge becomes apparent. Farming today is living on environmental credit – and the debt burden that is currently being built up will, at some point, have to be repaid.

With food provision, relatively subtle changes to the ecosystem could have a significant impact because many of the so-called 'renewable' resources of the planet are not infinite and must be rationed for there to be enough of them to go around. What is borrowed now (that which is consumed after the Earth Overshoot Day we identified in the Preface to this book) must be paid for at a later date, whether we like it or not. The so-called 'ecosystem services' embedded in our environment, that provide us with life-sustaining resources (such as food, water and clean air) in the present day, must be made good if they are to carry on doing so in a reliable manner for all concerned in the future. The limited nature of renewable resources does not have to mean there will be less food per capita to go round later, but it could. Everything depends on how human beings respond to the problem. Economic theory and history are there in spades to argue for feast, famine, or all shades in between.

If the credit crunch leads to an outright reduction in 'expensive' foods – expensive in both economic and environmental senses – then the red ink of financial debt will produce a green environmental result. The economics profession's most basic purpose is the allocation of finite resources among infinite desires. The world in the future may want a high-meat diet. Emerging markets may aspire to increase meat consumption (India being the notable exception, which – with around a sixth of the world's population – should not be ignored). However, the combination of rising relative prices for such environmentally costly product with more conservative budgeting from developed economy consumers is likely to result in a shift in spending patterns.

If meat becomes more expensive because the inputs into meat production become more costly, consumption of meat relative to other products will tend to fall. We saw just this event take place in China in 2007–2008. Meat prices rose (owing to disease in the pig population), which caused Chinese meat consumption to decline. None of this is to suggest a world of vegetarians – that seems unlikely. It is not even to suggest a return to the mid-19th century level of meat consumption. Perhaps the world economy will return to the 1960s – in terms of food fashions at least. It would take a shock to generate that, but a real income shock in the 1970s did lead to a shift in spending behaviour. A change in perceived living standards today could generate the same consequence.

The rationale for reducing meat consumption is that it is economically expensive, and more-constrained households will adjust the composition of their calorie intake to meet their perceived economic circumstances. However, cutting consumption is not the only solution. There is also the possibility of substitution.

Substituting

The economics

Consumers are clearly not passive in the face of budget constraints or increased price sensitivity. As well as trading down and shunning expensive foods, consumers may also choose to substitute one form of food for another. This is a trend long understood in the economics profession. Consumers are likely to embrace chicken or other similar cheaper alternatives if the price of beef becomes prohibitive (this necessitates frequent, automated adjustments to inflation rates to take account of shifts in consumption patterns).

If budget constraints in a more financial credit- and growth-restricted environment encourage a shift in consumption patterns towards cheaper product, we cannot simply assume that beef is abandoned in toto (or abandoned in favour of tofu). Market forces can simply shift from expensive beef to cheaper beef. Grass-fed cattle are shunned in favour of 'factory-farmed' beef. Organic free-range eggs are dropped from the shopping basket, to be replaced by other eggs. Organic egg spending in the UK fell 13 per cent in the first year of the financial credit crisis, and 24 per cent in 2009. Organic poultry and game sales by value fell by more than 28 per cent in 2009, though dairy was down only 5.5 per cent. Why is this? Organic product is priced at a premium to non-organic product. This premium reflects the economic costs associated with organic farming. Generally speaking, organic yields are lower than non-organic. To some extent, the premium may also reflect a marketing strategy on the part of retailers – organic is seen as a product that consumers are willing to pay more for. Consumers are motivated to pay the premium because they appreciate a difference in the flavour or because they perceive environmental benefits from organic farming (and are prepared to pay to support their principles).

Come an economic downturn, and the cost of those principles assumes a disproportionate weight in the mind of the consumer. The consumer still buys eggs (in the year to January 2010, as organic egg consumption slumped, free-range egg consumption *rose* 7 per cent), but chooses to 'trade down', from organic product to non-organic, cheaper product. Conscience has a price.

Higher-yielding agricultural production comes with a lower economic cost. At times of slower income growth and general household budget restraint, it is likely that the consumer will consider the trade-off between consumption and conscience.

This consumption shift is, it should be noted, the consumer's *choice*. Forcing a consumer to substitute cheaper product rarely works. The British government, facing a shortage of grain and rising bread prices, passed an Act of Parliament in 1800 insisting that bread had to be 24 hours old before it could be sold (hence the popular name, 'The Stale Bread Act'). This, it was thought, would discourage people from consuming warm, high-quality bread as a convenience food. It was as successful as one would suppose, and was followed in 1801 with the 'Brown Bread Act', forcing bakers to produce only the more grain-efficient wholemeal loaves. With callous disregard for their government's aim of grain efficiency, not to mention the health benefits of a high-roughage diet, the population rioted. The 'Brown Bread Act' lasted two months before it was repealed.

The environment

Substitution of cheap (high-yield) food for financially expensive (lower-yield) food has been the key feature of agriculture in modern times. However, the adoption of higher-yielding techniques in the name of cheap food provision has come at an environmental cost that is largely ignored in the market price. Cheap food is not cheap food.

Higher-yielding agricultural production has required an increased use of finite resources, generally speaking. Of course, these are resources that other sectors of the economy may be demanding. Some of the technology improvements in agriculture are (in effect) leveraging energy sources for which other sectors are competing, as well as creating a dependency of the food sector on finite resources. In leveraging manufactured resources, dependency on nature's cycles is potentially replaced by a dependency on the economic cycle.

Arguably, the more important issue relates to the human population cycle. Human ingenuity has (in effect) extended food supply by supplementing the renewable (but limited) food resources with other (finite) resources such as chemical-based fertilizers to increase the 'productivity' of the food and agriculture sector; however, this increased 'productivity' in the food and agriculture sector is not in fact an efficiency gain but, simply, an increase in resource intensity. The conclusion is an increase in finite resource use and when those resources are gone, they are gone. Even as far back as 1970, seven calories of non-food fuels were used to produce each calorie of food produced in the US.

The higher-yield agricultural practices that appeal to the cost-conscious consumer results in yield improvements that can survive *only* in the very short run. These improvements in yield are delivered by potentially compromising the security and quality of our food supplies in the medium term. Today's

Box 1.1 From gyrfalcons to Malthus

The biological sciences describe the interaction between food availability (or food yield) and population cycles in the animal world. Even the most powerful of species is not immune to this cycle. The gyrfalcon is the king of predators in the sub-Arctic, so much so that, historically, it was much sought after as a hunting bird in Europe in falconry circles. In its natural habitat, it preys on a small brown bird called the ptarmigan. Roughly every nine years, the ptarmigan population expands to the point where the available food supply cannot support it and population numbers then plunge. This cycle in turn regulates population numbers for the gyrfalcon, which are not (unlike human beings) endowed with the ability to see the peak of the food cycle coming. A gyrfalcon's young take about four years to reach maturity, so a rational gyrfalcon parent with a human ability to forecast population cycles would constrain its production of offspring at the peak of the ptarmigan cycle in the knowledge that, by the time chicks reached adulthood, food would be much more sparse (and, for the unluckiest gyrfalcon generation it would become even sparser through most of their reproductive life). In reality, of course, the opposite of rational happens: lush food supplies stimulate reproduction, with Malthusian consequences for the gyrfalcon every few generations.

food bill is not just charged to a financial credit card, it is being chalked up to an environmental credit card – which must be paid off at some point. As the medium-term costs are not reflected in the market price, the cost-conscious consumer is misled into consuming higher-yield product to maintain their short-term standards of living. The cost of the short-term gain – the ability to 'break' with the traditional 'seven years of fat and seven years of lean' – may potentially be that the human population cycle is tied to the much longer but 'one-off' process of resource depletion. Our food supplies (to state the obvious) need to be renewable. Any attempt to escape the natural cycle, or to move towards a high-yield cycle in the name of economic cost control, may in fact be to set up far greater Malthusian (and economic) problems for the future.

The economics of the dust bowl clearly demonstrates the problem of consumers being driven to high-yield agricultural produce derived from finite resources. New technology in the form of chemical fertilizers may at first sight appear to be a means of engineering a step-change in the productivity of agricultural land and allowing consumers to eat on the cheap. The evidence suggests that this is illusory. In the 1930s, new agricultural technology pushed American farmers to increase their field sizes in order to allow for a more efficient approach to field management (and thus higher yields). Over a period of half a dozen or so years, the removal of natural wind breaks together with extremely dry conditions resulted in a loss of topsoil from 100 million acres of farmland. The famous dust storm that hit Washington on Black Sunday, 14

April 1935, reportedly helped soil protection legislation (the Soil Conservation Act) pass in the same year.

The relative performances of the US and the UK agricultural sectors and the dangers of mistakenly seeking yield at any (environmental) price were understood by writers in the 1940s. UK yields were around double those of the North American prairies, but the (economic) cost of imported grain was far lower. Agricultural writers of the time declared 'we [the British] shall want to buy food for ourselves ... the reason for our buying should be that we cannot produce all we require or can afford to eat, *without extending production over land naturally unsuited to growing cereals or even grass*' (emphasis added, taken from Sykes, 1944, p13).

Improvements in agricultural productivity brought about by the so-called Green Revolution also illustrate the same point. The term refers to developments in agricultural technology in which a combination of techniques (plant breeding, irrigation and the application of agro-chemicals including pesticides) were applied in such a way as to enhance the productivity of the land. The first steps in the Green Revolution began in the middle of the last century when Mexico, faced with a growing population, set up an agricultural research centre under the auspices of Norman Borlaug with the aim of becoming self-sufficient. This goal was met by the mid-1960s. The same technology was gradually adopted by other countries, such as India, the Philippines and Africa, the main aim also being to help agricultural production keep pace with population growth.

The problem with the Green Revolution turned out to be fresh water. The US Geological Survey website documents a 300-foot decline in the level of groundwater in the aquifer that supports Santa Fe, New Mexico. In the US, the Ogallala Aquifer in the Great Plains feeds a number of states with their agricultural irrigation water supplies. Reportedly, 6 per cent of the aquifer has dropped to unusable levels.[6] The water issue goes well beyond agriculture because agriculture competes with other parts of the ('food') food-chain as well as with other industrial sectors. In India, Coca-Cola was accused of depleting groundwater even though the company website describes investments in rainwater harvesting in the country. The vexed topic of water is something we return to in the next chapter.

The Green Revolution increases yield and makes cheap food available for all. But it only makes cheap food available *for now*. The financial credit crunch is circumvented, but the environmental credit crunch is made more acute to bring about this outcome. If the Green Revolution is depleting natural freshwater sources, then the Revolution will turn to failure. As water resources fail, not only will the current agricultural yields be unattainable, but yields that were sustainably achievable in the past will be beyond the reach of future generations. Consumers' desire to maintain their consumption patterns by substituting high yield for low yield will eventually destroy their standard of living through the depletion of finite resources.

Technological advances can create the illusion of satisfying demand, but if they come courtesy of the depletion of non-renewable resources, the supply response is entirely illusionary. One thing that might help create more sustainable growth in the future is the fact that demand and consumption are not necessarily the same.

Waste not, want not

Walking towards a shopping mall in northern Chinese cities, it is not unusual to pass a barrow into which the locals throw food scraps. By the end of the day, this will be heaped high, and is then removed to feed local livestock; a highly effective approach to collaborative recycling. The parents of the baby boomer generation knew exactly the same process in English country villages: diets constrained by rationing could be supplemented by shared livestock, fed on shared scraps. Absolutely nothing was wasted.

During the First World War, food waste was a crime in the UK and incurred both fines and prison terms: 'it became a crime for a workman to leave a loaf behind on the kitchen shelf of the cottage from which he was moving (£2 fine) … for [a] lady in Wales to give meat to a St Bernard (£20), for a furnaceman dissatisfied with his dinner to throw chip potatoes on the fire (£10)' (Beveridge, 1928, p238).

Two generations on from the last wartime food rationing and the UK has become a society that throws away food with a casual disdain. A total of 15 per cent of *all* food and drink sold to households in the UK is not consumed, and is instead disposed of (one would hope composted). The total waste amounts to 16 per cent of calories consumed in the UK. The government response is to seek the end of the 'best before' date concept on food packaging (the consequences could be interesting as at least a generation has grown up not knowing how to tell if an egg is fresh – the solution is to place the egg in a bowl of water; if the egg lies flat on the bottom, it is fresh). In the US, 14 per cent of all food purchased (fresh or otherwise) is disposed of uneaten, accounting for 1.28lb per person per day. Fully 27 per cent of that waste (or 0.35lb) is vegetables. The British are healthier. They waste a mere 0.07lb of vegetables per person per day.[7]

To read a cookbook from the 1970s is to enter a different world. Delia Smith, revered by a generation of domestic cooks, dedicates an entire chapter of her *Complete Cookery Course* (published in 1978) to the treatment of 'leftovers'. Jamie Oliver, doyen of Generation Y cooking, devotes no pages whatsoever to the treatment of leftovers in *The Return of the Naked Chef* (published in 2000).

In an ideal world, food waste would at least be composted, but the infrastructure of urban living works creakily at best in this regard. Waste on this scale can be attributed to the changing nature of food consumption. The advance of affluence is normally accompanied by the attendant handmaidens of consumer durable goods. Fridges, freezers and microwaves allow consumers to consume food in a substantially different way from that of a generation ago. Longer-term storage allows people to keep food for longer, but also to forget about the food that they have stored. Unearthing a freezer-burned box of some vaguely turkey-based product a year beyond their sell-by date should (one hopes) prompt the discerning consumer to throw the whole thing into the nearest bin.

Longer-term storage also allowed a shift in consumption spending to occur. Daily shopping was a routine for a society with low female participation in the workforce, and a limited ability to store food for long. However, purchasing on a daily basis reduced the amount of waste. For one thing, food was bought with a specific objective in mind. For another, bulk buying was discouraged if the purchases had to be carried around a series of smaller shops. As female participation in the workforce has risen and society has (collectively) become time poor, so there has been a shift to lower-frequency purchases. The family now embarks on a weekly shop, rather than a daily shop; in some cases even less frequent than weekly shop visits.

This move to supermarket based, but less frequent, shopping visits has a number of environmental consequences. There is an increase in car use (relative to the daily stroll to the local shop of yesteryear). However, given that society has changed the way it views transport (at least for the time being), the modern alternative to supermarket shopping is probably not a pastoral ideal of a gentle stroll to a selection of village shops, pulling a shopping trolley behind one. For one thing, the local village shops no longer exist. High-frequency purchasing would likely generate more car use.

Supermarket shopping also encourages impulse buying. This phenomenon was first remarked on in the 1950s, when the supermarket was introduced in the US. Observers commented on the 'trance like' state of shoppers as they moved around the supermarket aisles.[8] Initially, as the shoppers made the transition from daily shopping in a variety of stores to the new supermarket experience, they found that they arrived at the checkout with insufficient funds for the purchases that they had made – because they had bought products that they had no intention of consuming. Cynicism and practice makes this less likely nowadays, but the phenomenon of encouraging additional and 'unnecessary' purchases is a powerful force in the supermarket industry. The BOGOF (buy one get one free) concept is one of the more blatant attempts to increase consumption through impulse. It is also a sales technique that explicitly depends on waste to be commercially viable (unless it is a designated loss leader). The supermarket makes money if the consumer undertakes additional 'reckless' or unnecessary consumption of the BOGOF product, or if the second product is 'past its best' before the first product has been completely consumed.

Finally, the modern consumer is not, generally, that good at planning ahead in terms of food consumption. A future meal of grilled fish and salad

Box 1.2 Social enterprise and the local shop

Although size can bring benefits (such as 'cheap food', in economic terms), scale comes with intangible but nevertheless very real costs. The concept of 'social enterprise' is relevant in the debate about the infrastructure that allows us to consume food. A social enterprise is a business run in an efficient way from a cash perspective, but which is also valued for its intrinsic social usefulness. The stakeholders in such a business can be paid 'in kind' by the delivery of useful services that might otherwise be lost in the race for high yield, market domination and shareholder returns. The UK radio drama *The Archers* highlighted this with a storyline about the creation of a community store. Faced with the disappearance of a local shop (as supermarket competition increased), local villagers chose to buy the business and run it as a community. Of course, such enterprises (whether in fiction or in real life) must be commercially viable or fail. In the final analysis, however, the community store exists for the local community and not for the shareholder. Any 'value added' is ploughed back into the business. In this way, the position has parallels to the very (economically) powerful cooperative store movements of the 19th and early 20th centuries.

with a bottle of wine may seem a great idea when purchased three days in advance. The weary economist, returning from a day of struggling to understand the latest complexities of global monetary policy, may simply lack the energy to prepare the food – and instead call for a takeaway pizza while disposing of the raw ingredients (the wine, it should be observed, is likely to be salvaged in this process).[9]

The economics

So, how does the new consumer environment change the pattern of food waste? There are two strains to the shift in consumer behaviour, which are not necessarily complementary.

First, we return to the disproportionate price sensitivity of the consumer to food prices. Food waste is a visible 'money down the drain' issue. Consumers who feel budget constrained in a financial credit crunch are more likely to seek to control their food budget than other forms of spending. This is likely to come through in the form of more cautious and planned food shopping – the Sunday newspapers' magazine supplements were full of advice on how to write a shopping list as they hyped up the fear of food price hyperinflation in 2008.

It is very clearly noticeable that food waste in the UK is strongly skewed towards the less expensive items of the shopping basket. Consumers are already price-sensitive when it comes to the issue of waste. It seems reasonable to suppose that elevated price-sensitivity and constrained incomes will change behaviour among those with an extraordinary propensity to waste food, combining to create a more economically and thus more environmentally restrained food consumption basket.

Against this, however, is the second reaction of a household confronted with credit constraints and slower income growth; to increase household income through maximizing household earning capacity. Anglo Saxon economies have already seen an increase in female participation in the workforce. Japan's female workforce participation rose steadily during the economic 'lost decade' of the 1990s, as households sought to maintain their standards of living with an additional income. Households in Europe are also seeing an increase in female participation. If credit is less freely available, or income is more constrained, the reaction of the consumer seems to be for one-income households to become two-income households (or at least to become one-and-a-half income households, through the medium of part-time work).

This raises a challenge in attempts to change consumption patterns so as to reduce food waste. Food preparation in a society whose households are increasingly 'time poor' is likely to favour convenience food. They are also likely to favour shopping less frequently, not more frequently. Both of these consequences have implications for the environment. The former increases energy consumption: both in the preparation of food for retail and in the reheating of food for home consumption (TV dinners reheated in an oven are energy intensive forms of preparing food). Convenience food also comes heavily packaged, which also incurs environmental costs.

The environment

The level of food waste in developed economies is a failure of the market. Food that is wasted in developed economies could be a (literal) lifeline to consumers in other economies. The US population is wasting 36 *billion* calories every day, in vegetables alone.[10] To put that in context, US vegetable waste alone is more than 26 per cent of southern Africa's vegetable supply (by calorific content).

This is more than a straightforward credit concern. American consumers (and consumers elsewhere in the developed world) are borrowing from the environmental future, and then wantonly destroying part of the proceeds. It is like taking out a mortgage from the bank, and then burning down the top storey of the house.

Cutting out waste does present another challenge, however. If the US wastes less food, for the rest of the world to benefit then that food must be traded. The issue of trade in food – and the distances food has to travel before it is consumed – remains at the heart of the post financial credit crunch consumer environment.

Food miles

The 'food miles' issue has continued to loom relatively large in the minds of consumers. Consumers' perception of planes flying in product that could be grown locally is an obvious environmental target. However, the issue of food miles is not necessarily quite as critical a problem as the popular perception implies. Transportation accounts for 11 per cent of the fossil fuel consumption of US food production, distribution and preparation. To be sure, this is not a negligible amount, but it is dwarfed by the fossil fuel consumption involved in processing and packaging food (45.6 per cent of fuel consumption).

The economics

The price of food represents the inputs, of course. If those inputs reflect some of the environmental consequences, then the shift in consumption patterns arising from changing budget constraints can yield 'unintended' environmental benefits. Let us consider the tomato. Tomatoes can, of course, be grown perfectly well in the UK. Nevertheless, the UK still flies in tomatoes from Spain. The UK consumed 180,949 tonnes of Spanish tomato imports in 2005[11] – an even 3kg of Spanish tomatoes per person. Why? Because Spanish tomatoes are cheaper, in economic terms. Of course, Spanish tomatoes may be cheaper in economic terms because the environmental costs of flying them to the UK are not properly reflected in the price. But, as we shall see, the environmental costs of growing tomatoes in the UK happen to be higher than the environmental costs of flying them in. So, in this case, we appear to have an alignment of economic and environmental costs.

The environment

If a consumer is looking to buy a tomato in the winter, then the Spanish option is likely to be more economically and environmentally friendly. Tomatoes tend not to thrive in the British winter climate. As a result, they have to be raised in heated greenhouses, requiring energy to keep the ambient temperature correct. In Spain, however, there are fewer problems with the winter temperature, and tomatoes require less cosseting. Of course, there is an economic and environmental cost involved in flying the tomatoes to the UK, and distributing them on from their point of landing, but does this cost outweigh the heating costs for domestically produced tomatoes?

Generally speaking, the Spanish tomato (in winter) requires less fuel to get it from pip to plate than does a British tomato. With fuel prices likely to continue to rise relative to incomes in a more sclerotic (developed economy) growth environment, this means that the fuel component of a tomato's costs will rise relative to consumers' income levels. Thus, price-sensitive consumers will shun those foods that consume more fuel.

On one level this is a simple (market-driven) resolution to the debate about the most environmentally friendly method of consuming food. If any foodstuff accurately reflects the price of its components, and those components accurately reflect the environmental costs of their consumption, then those products that do most environmental damage will see reduced demand as prices rise. The beauty of this process is that pricing induces an environmentally beneficial response on the part of the consumer, without any need to educate the consumer on the environmental consequences of their spending habits. The pricing mechanism does, generally speaking, work.

Economics fails the environment when the environmental costs are not fully reflected in the market price. It sounds crazy to fly shrimp from Europe to Thailand for shelling, only to fly them back again for consumption. However, if the market price for jet fuel is sufficiently low, then the economics of this may work. This then raises the question of whether the market cost of jet fuel is accurately taking into account all the environmental consequences – including the impact on future generations as the environmental credit crunch applies even more pressure.

This is not to suggest that trade should be reduced *tout court*. However, the balance of short-term and long-term considerations needs to be factored in to market prices. Wasting food can be considered a depletion of constrained environmental resources across international borders, because of the way that food is processed. Should this lead to hard real effects – feeding some countries by putting others at risk of food scarcity – this must inevitably raise the spectre of geopolitical risk. Consider those countries buying agricultural land elsewhere in the world: in the depths of an environmental credit crunch, who exactly would eat the produce from that land? Those willing and able to pay (the landowners), or those in need of the food to stay alive? Famine tends to be thought of as a country-specific issue but, in the context of globalization, the causes of famine can be considered to be globalized. The spectre of starving children in the media has historically done little to change Western habits, and food aid has tended to be the response to famine in the poorest of countries. Ultimately, issues of food security may lead to countries or regions attempting to become more self-sufficient (which may actually increase the risk of famine in some parts of the world and which certainly has other environmental implications). In the meantime, it is more likely that economic forces will effect some of the changes, but probably only for the developed nations.

The financial credit crunch, the environmental credit crunch, and food

The economics of the financial credit crunch will change the way we consume food. Because this recession (globally) counts as the worst of the post-war cycle, and because the recession is also likely to engender a lower trend rate of growth in the future, we are expecting the shifts in consumption to be greater than in previous economic downturns. However, the changes that the financial credit crunch effects are unlikely to be sufficient to resolve the impending environmental credit crunch on their own. Even after the financial credit crunch, the global consumer will still be borrowing at an unsustainable rate. Rather than cash, the borrowing is from future stocks of finite resources. In other words, the financial credit crunch on its own is unlikely to push the Earth Overshoot Day to 31 December, though it should push the date somewhat later into the year.

What else is needed?

Behavioural change is a necessary part of the process. One of the reasons that the financial credit crunch cannot fully redress environmental concerns is that market pricing does not necessarily fully capture environmental costs. If the environmental cost is fully reflected in the price, credit-constrained consumers will be more environmentally friendly. If the environmental consequences are not fully reflected in the price – but instead form a debt to be paid in the future – then the financial credit crunch will not impact the environmental 'credit bubble' of consuming now without regard to future consequences.

The environmental 'credit bubble' that precedes the inevitable environmental credit crunch is probably best known under another name – the Tragedy of the Commons. This refers to the overexploitation of property in common ownership, for example common land. This was expounded by the economist G. Hardin more than a generation ago. Hardin predicted that, in the absence of ownership or regulation imposed by outside bodies, human beings will behave as (apparently) irrationally as a gyrfalcon. For an individual herdsman on common land, the only rational course of action (as an individual) is to add as many animals as possible to his herd. The gain from increased output accrues entirely to him on an individual basis, and the environmental cost of doing so (the effects of overgrazing) is shared by all the herdsmen. The environmental cost (overgrazing) is not the same as the economic cost. The consequence is a depletion of natural resources to exhaustion.

Hardin concluded that in the presence of increasingly large population numbers, the 'commons' must be abandoned. Indeed, in some circumstances – the enclosure of farm lands, and restriction of pastures and hunting and fishing areas, as well as restrictions on the disposal of human waste – this has happened. (Hardin also suggested that many other 'commons' would need to be abandoned, including the freedom to have families of unlimited size). In contrast, Elinor Ostrom received the Nobel Prize in 2009 for more optimistic work. She argued that subject to certain conditions – rules evolved over time, conflict resolution measures and an individual's duty to maintain common resources in proportion to the benefits from exploiting it – people can manage natural resources effectively without government intervention. In aggregate, these theories suggest that human beings could potentially manage the balance between population growth and food successfully but that it is unlikely to be very easy. Moreover, it requires behavioural change from the status quo.

Historically speaking, technology has come to the rescue through agricultural productivity, which has met increased demands without the need for consumer behaviour to change. Today, however, new technology may just inflate the global climate credit bubble further – solving one problem only to bring along another. Not infrequently, when outcomes are unexpectedly adverse, the issue turns out to be rooted in an aspect of 'commons' that is either unprotected because it is not owned, or is unprotected because its owner has no incentive to protect it. Technology developments in particular raise interesting questions about the possibility that some global resources may end up in the wrong hands.

From the perspective of environmental impacts, in particular biodiversity, so-called 'GM' can be regarded as an opportunity or threat. The 'opportunity' is for this industry to deliver seeds that are immune to salt, drought, flooding and pests. The 'threat' relates to the impact of the introduction of 'synthetic' species on the broader environment. Agro-chemical firms appear to be patenting traits as fast as possible. From Hardin's perspective, this may turn out to be a version of the Tragedy of the Commons. The question is whether the private sector should own genetic traits, or the public sector. Experience to date suggests that 'biodiversity' may be a 'commons' that is not sufficiently protected. On the one hand, this suggests that concerns expressed in the press in recent years by environmentalists in relation to the 'great extinction event' underway are real. On the other, it highlights a risk inherent in agricultural practices. Modern agriculture is based on monocultures, which improves the consistency of harvests from one year to the next; in this sense improving food security. This approach fits the 'factory' approach increasingly applied in order to deliver cheap food in modern times.

However, the monoculture, because it reduces diversification, may also increase the risk of extreme events. The Irish potato famine came about for multiple reasons but monoculture appears to have been a contributory factor to the disaster. In modern times, the disappearance of significant numbers of species from the animal kingdom is not well understood but one hypothesis is that agricultural practices based on monocultures may be to blame, along with other accoutrements of modern agriculture. In the early years of the 21st century, for instance, a significant fall in the population numbers of that all-important pollinator, the honeybee, has been widely documented. Beekeepers from the US to the UK and beyond have reported colony collapses and, although a number of causes are under discussion, this too is not well understood. The bee story may be capturing the attention of the population at large rather more than the 'fish stocks' story does because of the potentially sweeping effects on agriculture should the crash in bee populations prove to be permanent. A very significant part of the entire agricultural food chain rests upon successful pollination, and this translates to significant amounts of money. In the US, some US$14 billion in agricultural crops is reportedly dependent upon pollination by bees. The all-important point is that the reason for the disappearance of the pollinator is not well understood although some suspect it may be the combined impact of aspects of modern farming methods.

The most important point about the technology-driven failures described above is that it is not the technology itself that is the issue but the way it is used. The economics of the farming sector do not respect (cannot be expected to respect) valuable 'commons' unless these 'commons' are overtly embedded in the market pricing structure driving the agricultural industry.

Conclusions

The food story is not purely about numbers of people. Pressure on the agricultural sector to deliver cheap food for the population at large, particularly in wealthy countries, comes not only from population growth but also from the way food is consumed. As a consequence of these changes, manufacturing processes (and content) have also been widely adopted in the food processing industries. The once-fragmented food processing and distribution industries have become dominated by large players. In becoming large they have brought about big changes in the way food is procured, transported, processed, merchandized, prepared and consumed, usually in the name of a narrow economic definition of 'efficiency'. (Such 'efficiency' addresses the need to provide cheap food, discussed above, as well as the constraints typical of an economic credit crunch.) The number of calories and grams of protein consumed per head per day tends to rise with wealth, along with greater variety in consumers' diet. New modes of travel and communication have accelerated these trends globally. In the past two or three decades, in some societies, people have also become increasingly time poor. These multi-dimensional changes have combined to lead to a transformation, not only in *what* is consumed but also *how* consumption happens in wealthy countries. This issue has been recognized by some consumers already, hence, growth in the 'organic' and 'slow food' movements observed over the past five to ten years.

The financial credit crunch is changing how food is consumed, certainly. The immediate effect is probably helpful in diminishing the impact of the environmental credit crunch (pushing out the Earth Overshoot Day). However, the impact of the financial credit crunch is not all one way. In particular, policymakers may seek to perpetuate the low-cost rationale in the short term (to mitigate the effects of the financial credit crunch) without a thought for the long-term environmental credit crunch consequences. The fact that this feeds back into economic implications is generally ignored – as the cost (economic and environmental) is beyond the electoral cycle.

Whether market forces, in the wake of the financial credit crunch, produce a truly rational usage of 'natural' food stocks will depend substantially on the institutional structures (governance, trade law, market practices and the like) within which those forces are constrained. To date, the financial credit crunch has had only a limited positive or negative impact on the approaching environmental credit crunch for food.

The additional ingredient in the food debate is, therefore, behavioural change. The question is what paradigm shift will address the problem of numbers. There is a danger that, in spite of clear expectations of a dramatic population rise by the year 2050, governments fail to respond and fall prey to an agricultural environmental credit crunch. Any famine that should come to pass in the middle of the century can be called environmental but, in this case, could also be called policy induced. As with the financial credit crunch, policy error or inaction early in the process has the potential to produce calamitous effects in the denouement of the crisis.

Notes

1 Cited in Sykes (1944).
2 In the US Congress, 700 secretaries protested to the Speaker about food price rises in the cafeteria of the House of Representatives – a fact that the Speaker made public, in an overt attack on the President's policies (see Bowles, 2005).
3 The data is calculated from the chain weighted real consumer spending on food and beverages, and on restaurants and cafés, taken from the detail of the quarterly UK National Accounts data.
4 The data on energy usage is from Rich Pirog, cited in McWilliams (2009, p25). Food spending in restaurants is around 40 per cent of an American's total food budget (from the breakdown of consumer spending data in the National Accounts) – which is in line with the energy consumption of restaurants. However, the retail cost of the food used in restaurants is around 20–35 per cent of the final sale price. The mark-up is used, of course, to finance labour costs, other overheads and the restaurateur's profit.
5 Calculations are derived from weight of food consumed. The workhouse in 1833 is drawn from the statistics relating to the Gosport Workhouse cited in *The Westminster Review* (1833, p465). The British soldier in 1833 is from the same source (p469). British wartime rations are sourced from www.wartimememories.co.uk/rationing.html. The world aggregates come from UN Food and Agriculture Organization (FAO) data cited in Woudhuysen and Kaplinsky (2009, p334). Conversion rates for translating food weights into calories, and the calorie content of the quarter-pounder meal, come from www.freedieting.com.
6 Details of this estimate are sourced from http://academic.evergreen.edu/g/grossmaz/WORMKA.
7 The statistics on British food waste come from the Department for Environment, Food and Rural Affairs (Defra), at www.defra.gov.uk. US data is cited in McWilliams (2009, p28).

8 The social revolution that was created by the arrival of the supermarket is documented in Packard (1961).
9 According to Defra, 6 per cent of alcoholic drinks are wasted (versus 15 per cent of all food and drink). Some might express some surprise at so high a percentage.
10 Clearly, Americans are wasting vastly more calories than cited here, through non-vegetable food waste. However, without knowing the precise composition of what else is being lost, it is difficult to calculate the exact loss in calorie terms.
11 Data here is taken from the UN FAO databases.

References

Beveridge, W. H. (1928) *British Food Control*, Oxford University Press, Oxford
Bowles, N. (2005) *Nixon's Business*, Texas A&M University Press, College Station, TX
McWilliams, J. E. (2009) *Just Food*, Little, Brown & Company, New York
Packard, V. (1961) *The Hidden Persuaders*, Penguin Books, London
Russel, A. (1864) *The Salmon*, Edmonston and Douglas, Edinburgh
Sykes, F. (1944) *This Farming Business*, Faber & Faber, London
The Westminster Review (1833) 'Poor laws commission', April, pp465, 469
Woudhuysen, J. and Kaplinsky, J. (2009) *Energise*, Beautiful Books Ltd, London

CHAPTER 2

Water in the Wake of the Two Credit Crunches

Nothing in the world is more flexible and yielding than water. Yet when it attacks the firm and the strong, none can withstand it, because they have no way to change it. (Lao Tzu)

It is fortunate that, as well as being a necessity of life for human beings, water is also the world's ultimate recyclable resource. The water in the cup of tea that the discerning reader may be sipping while reading this could previously have been part of Queen Victoria's bathwater, have doused fires in London in 1666, and been a component of the ice sheets that covered the UK and continental Europe 18 millennia ago. Economics, which is concerned with the allocation of finite resources among infinite desires, surely has little to say on the subject of a resource that is, for all practical purposes, infinite and constantly recycled? Of course, economists are never at a loss for things to say – and the problem of water is not that the supply of H_2O is finite; the problem is that the supply of usable water (including, specifically, drinking or potable water) is finite in practical terms.

Every aspect of human activity – food and drink, dwelling space, clothing, infrastructure, urban environment, manufacturing, housekeeping, transport, health, leisure and culture – has been, from the dawn of civilization, connected to water. However much economics may have to say on the subject of resource allocation, the abstract world of economic theory immediately runs into a real world problem. Water is not like other commodities. A human can survive for between two and ten days without water, but that is all. Even modest dehydration will deplete a human's efficiency. After oxygen, water is the most important element for survival.

Thus, at the extreme, access to water can be described as a human right because it connects directly to the right to life. For so critical a resource, market economics can play a very limited role. However, water is also a necessity of economic life and that means economics must come into the picture somewhere. Most of the major cities of the world are sited on a river, river delta, or a large body of water, because this made strategic sense – a perspective that of course included economic considerations such as the need to trade or to irrigate food crops or to move people or goods from one place to another.

From ancient times to the modern day, the river continues to provide water for domestic and industrial use, water for agriculture, as well as infrastructure for transport.

When the natural infrastructure is used *in situ* (avoiding the need to build canals, reservoirs, irrigation feeds or other conduits, thereby avoiding significant economic costs), the water contained in a river or lake takes on two dimensions: it is integral to the ecosystem and yet also integral to the urban infrastructure and human needs it feeds. Politics is an inevitable part of the picture because the distribution of usable water around the world is uneven, and even though many population centres initially grew up around water supplies, the allocation of abundant water around the globe does not necessarily correspond to the major centres of population in the modern age.

These, then, are the three points that encapsulate the central dilemma of this chapter:

- Water is at one and the same time a necessity of life and commodity in infinite supply;
- Water is an important 'commons' (the concept we introduced in Chapter 1) in the form of a key ecosystem service, but it is also an economic good; and
- Water is everywhere but unevenly distributed.

Water is a socio-economic hybrid. Sometimes it is clear that decisions governing its use cannot and should not be subjected to economic considerations, because the best outcome must be driven by values. At other times, it is clear that a common sense application of economics might be very constructive in better balancing supply and demand in the context of supply constraints. However, many situations are ambiguous. This may well be why no less a body than the United Nations Conference on Environment and Development (UNCED) declared in the 'Dublin Statement' in 1992 that 'Water has an economic value in all its competing uses, and should be recognized as an economic good.'

Unsurprisingly, the UNCED statement has proved to be highly controversial among environmentalists and human rights organizations. However, if economics does have some role in the consumption of water, and if the economic landscape is changing as a result of the financial credit crunch, then an analysis of the financial credit crunch must consider the impact on water.

Drinking water will ultimately be demanded at any price given that it is a necessary condition for life; and sanitation and clean drinking water connect directly to human rights. In an ideal world, economics would come into play at the right moment and step back when values should drive decisions. This is far easier said than done in the context of water. First, human life is priceless; so, therefore, is the water that supports life. The market is poorly equipped to deal with infinite prices. Second, one problem inherent in 'commons' is that the forces of economics are unable to balance supply and demand – in effect, there are no brakes until the relevant resource runs out. Third, water can be a finite resource in specific geographies and will tend to run up against infinite needs. The nature of water means that it does not respect human boundaries – so water disputes easily arise between different political states. Managing a river in one political state (so that it is appropriate for the local population) may alter the supply in a neighbouring state (creating a finite resource where previously water was assumed to be an infinite resource).

It seems an absurd understatement to say that the issues surrounding water are complex. Breaking the issue down, let us start with first principles and look at the way in which water flows into human society – the hydrological cycle.

The hydrological cycle

In some respects, the hydrological cycle is rather like the economic cycle. Water evaporates from the surface of land and water, becomes water vapour, and returns to the Earth as rainfall; this is part of the 'flow' in the hydrological cycle (the equivalent of economic gross domestic product (GDP) for the water environment). Some of this goes to replenish the stocks of water held beneath porous rock – groundwater in the metaphorical savings bank (wealth, or the stock of assets, to adopt the terminology of economics). Once underground, it can stay there for a very long time unless it is drawn down by human beings (reducing the stock of savings – as happened in monetary terms during the run-up to the financial credit crunch). Groundwater sometimes flows, albeit extremely slowly, so some groundwater stocks are known as 'fossil' groundwater.

Meanwhile, some rain hits fertile soil and generates plant growth. This process of moving water through plants (transpiration) is known as 'green' water. Some water filters through rock, and some runs across the surface, to enter lakes and rivers, and a portion of that ends up in the sea. The water held in rivers and reservoirs is 'blue' water.

Just as, in economics, it makes sense to live off economic flows (rather than drawing down excessively on stocks) so, when it comes to the hydrological cycle, it makes sense to use the water held in 'banks' such as aquifers sparingly, and shorter-term flows (especially rainwater) more freely. Even flows can be overused, however: if too much rainwater is harvested (or even simply prevented from soaking into the ground by being channelled down the urban drain system), this can prevent aquifers from filling up, inadvertently depleting long-term water banks. It is the same with economics – if all income (flow) is consumed and nothing is saved then the value of savings might diminish (eaten away by inflation, taxation or both).

Drawing on aquifers is a perfect instance of a potential environmental credit crunch, of course. Consuming water in this manner today *has* to be repaid with less consumption of water in the future.

A sense of proportion

So, if it makes sense to live off flows and not hasten the environmental credit crunch through the consumption of stocks, how much water are we talking about? Around 70 per cent of the world's surface is liquid water, 5 per cent is covered by ice. However, very little of this water is actually supplied to humanity (or land-based nature, for that matter). In fact, only around 0.1 per cent of the world's water is available for consumption from water flow.[1]

Much of the world's water is held as sea water, which is not (without treatment) usable. Of the world's freshwater supplies, 60 per cent is trapped in the ice of the Antarctic. A further 6 per cent is in the ice sheets of Greenland. Around 30 per cent is groundwater, or fossil water (in aquifers and rock).

Every year, 113,500km^3 of water falls to Earth as precipitation. This is the aforementioned 0.1 per cent of the total water on the planet. Only 0.1 per cent of the water on the planet is available for use by humans *and* land-based nature combined. This small a percentage seems too derisory a quantity to be plausible (particularly from the vantage point of a rain-soaked UK). However, with a global population of 6.5 billion, 0.1 per cent of global water stocks falling as precipitation can provide enough water to supply every person with 320 full bathtubs of water. That is 320 bathtubs of water per person, *every day*. Even with United Nations (UN) projections that the world's population will hit 9 billion by 2050, there will still be 231 bathtubs of water per person per day. Bear in mind that this is precipitation – the natural water cycle. There is no consideration for the use of so-called 'grey' water – using bathwater to irrigate the garden, for instance. (The precipitation statistics come from Mauser, 2008, converted into bathtubs by the authors.)

The finite part of water supply is therefore sizeable. However, if that water is not available in a geographic region (too little rain) then some kind of economic solution will need to be found.

Summoning rainmakers – the ability of humans to change the hydrological cycle

It should also be borne in mind that the hydrological cycle is impossible to separate from human activity; in fact, human activity can affect the hydrological cycle in two ways: directly, for example, by overusing natural water banks (aquifers), or by polluting open water or the natural infrastructures that contain and filter it with the detritus of economic activity; and indirectly, by changing the ecosystem's natural infrastructure to such an extent that the hydrological cycle starts behaving differently.

The most extreme (and high-profile) version of this is man-made climate change, but there are plenty of smaller-scale examples. In Mexico, the capital city is sinking because the groundwater beneath it has been overdrawn, causing soil subsidence. This is not simply a question of inconvenience (cracked buildings), but also becomes an issue for freshwater provision, as sewage pipes crack, leaking human waste into the city's drinking water supply.[2]

In western and southeastern Australia, deforestation resulted in a rise in the water table, and as a consequence the topsoil became salty; this change to the water table may be permanent without significant efforts to replant the trees, and the longer-term consequence may be the replacement of a fertile part of the ecosystem with a new area of desert. In some river deltas, formerly freshwater aquifers are now brackish because, when the dams stop the river from flowing to the sea, seawater invades the delta; one of many examples worldwide was the state of the San Joaquin/San Francisco Delta in the mid-1990s, where river flow was said to be reduced by 90 per cent, reportedly costing the local fishing industry US$3 billion.[3]

River basins provide human beings with water for irrigation, drinking, waste disposal, transport and leisure. Inevitably, some human activities compete with others. Whereas transport can go on almost undisturbed no matter what the state of the water, too much use for waste disposal compromises or destroys any food stocks the river might carry, and of course makes it more challenging to use the river system for drinking water. As this suggests, human activity can also have a significant environmental impact on river systems; there may well be almost no large river system in which the hydrology is unaffected by human activity, except (perhaps) in the most sparsely populated regions.

> *Within large river basins many factors can simultaneously affect runoff, such as abstractions for irrigation and other agricultural purposes, as well as for industrial and municipal water supply. There can be soil drainage, deforestation and afforestation, agrosilviculture, urbanization, open-cast mining and mine water pumping, stream bank straightening, and excavation of sand and gravel from river channels as well as other activities. There may be large-scale diversions of flow from one basin to another and river flow control by reservoir operation.* (Shiklomanov and Rodda, 2003, p27)

Consider just a handful of the many press headlines relating to the state of the world's rivers. The Rio Grande is described as 'sucked dry'. In Hyderabad, India, an 'Indus River Day' ceremony was held over the 'dry riverbed' of the 'once mighty' river. In 2006, China's Yellow River reportedly turned red on several occasions, by accident, not by design. However, governments can and do respond to the risk posed by environmental degradation – in the US, the Environmental Protection Agency (EPA) has an established programme for the restoration of water-based ecosystems such as wetlands and watersheds. Moreover, the European Water Framework Directive is moving slowly towards an integrated system of regulation for river basins.

Of course, water has long been a source of political friction – witness Spain and Portugal, Singapore and Malaysia, or China and India in the present

Box 2.1 The Aral Sea

The Aral Sea demonstrates the many problems of economic interaction with water. The draining of the Aral Sea started in 1918, when the USSR sought to irrigate the surrounding land to cultivate 'white gold' (cotton). Rivers feeding into the sea were diverted. The rationale for this was an undervaluation of the economic and environmental value of water – the USSR knew that its action would lead to the draining of the lake, and indeed it was surprised that it was not happening faster.

After the fall of the Soviet Union, the Aral Sea was divided between Uzbekistan and Kazakhstan. By this stage, the sea had divided into two, and pollution levels rose dramatically. Pollutants became concentrated in the two lakes. The irrigation systems were not efficient; a high proportion of the diverted water never reached the cotton fields that were the objective of the irrigation schemes.

Bad economics, in mispricing water, essentially created the problems of the Aral Sea. The role of economics in environmental solutions has also been highlighted by subsequent developments. Kazakhstan has oil wealth, which allowed the Kazakh government to invest in its infrastructure. The improvements meant that the northern half of the Aral Sea began to recover, and its water levels rose. The Uzbeks, however, lacked the wealth necessary to finance investment. Only 12 per cent of the Uzbek irrigation canals are leak-proof. The southern portion of the Aral Sea fed by Uzbek water sources continued to deteriorate. The response of the Kazakh parliament in 2003 was to sanction the building of a dam, to prevent 'its' water in the northern sea flowing into the southern. The dam was completed in 2005, and within less than three years the water level in the northern sea had risen from 30m to 38m. Experts consider 42m to be the level at which the northern Aral Sea would have a viable future. The southern Aral Sea, meanwhile, continues to deteriorate.

The Aral Sea is a natural monument to the problems of economics interacting with water, and the damage that a lack of capital can generate in water management.

day. Water has also been a cause of warfare (Darfur) as well as an instrument of war. Rivers have often provided a natural defence, and many famous battles have featured the destruction of dams or bridges to deprive a city of access to key resources.

Marco Polo is said to have stood at the gate of the Black City on the old Silk Road watching a bustling city of 3000 souls. To stand at the same gate now is to see crumbling walls slowly being swallowed by the desert. The modern tourist might wonder how this city survived in such an arid place. The answer is that it has not always been so dry, the Black River once ran through this city. This reliable source of water made the city impregnable in times of war, until the

Han Chinese, besieging the city without success, hit upon the idea of blocking the river. The inhabitants of the city inevitably had to give in and charged through a hole in the wall in order to go down fighting. The Han Chinese swaggered into the city, but their victory literally turned to dust. The course of the Black River had permanently changed in response to human intervention, and human intervention was powerless to reverse the change, leaving the city permanently without water. The spoils of war slowly dissolved into the sands.

This salutary example illustrates the extent to which the natural infrastructure that delivers freshwater resources is ill-understood, and how economic or any other activity undertaken without an awareness of environmental costs can have unexpected consequences. Science has advanced, but this still tends to be true today because the ecosystem falls into several branches of science – what is really needed is a systems science approach in which several disciplines can be combined. When water is abundant it is unlikely to become political, but the fact is the distribution of usable water around the world is uneven, and scarcity is increasing, which will tend to crank up geopolitical pressure. So, what does 'increasing scarcity' mean? The established way of describing the state of water availability is defined by the UN. A country or region is described as suffering from water stress when available water per person per year is 1700m^3 or less, equivalent to a maximum of 31 bathtubs of water per person per day; water is said to be scarce when there is less than 1000m^3 per person, equivalent to 18 bathtubs per person per day; and water is said to be absolutely scarce when there is less than 500m^3 per person – nine bathtubs per person per day. This may sound like plenty of water in the context of household demand. Households based in countries already facing sub-nine bathtubs per person per day availability typically use only around 2 per cent of what water is available for household needs (washing, cooking, drinking and waste management).

This, however, misses three key points. First, water is not evenly distributed through individual countries; in countries with access to nine bathtubs per person per day on average there will be many households with access to far less than that. Second, an increasing number of countries are moving into the 'water stressed' or 'water scarce' categories. The United Nations Environment Programme (UNEP) suggests that in 2025, not only will a greater number of African countries face absolute water scarcity (nine bathtubs per person per day or less) than currently, but significant areas of the US and Asia will fall into the sub-31 bathtub zone. About 6 per cent of those 31 bathtubs would be required by the average US household for domestic needs.

If this were the whole story there would be no problem. Unfortunately, it is not. The third and most important point is that the household usage numbers just cited do not reflect the actual water *footprint* per household because they do not include the water needed to provide the household with food.

The question is how many hamburgers per year could be afforded by the sub-31 bathtubs per person fast-food meat-eating household? The answer is to be found below – and hamburger-eating readers may want to prepare themselves for some bad news. The way water is used by significant numbers of people will have to change as shortages bite home. If it were simply a matter of consuming fewer hamburgers per year, this would presumably not be too much of a sacrifice and might even bring health benefits. However, people in some countries will face harder choices – including whether to stay in a region that does not have enough water to grow food crops, and whether to export food to other countries without reflecting the use of this scarce resource in the export price.

Water and the twin credit crunches

So, where does this leave the economics of water and, specifically, the economics of water in the wake of the financial credit crisis? Economics interacts with water in four specific ways:

- Determining the allocation of water resources between various demands (personal, agricultural and industrial consumption);
- Funding the infrastructure that allows for efficient use of water;
- Creating incentives for water efficiency; and
- Virtual trade of water.

Economics can ease or hinder the efficient use of water through these four areas. As the financial credit crunch changes the economic landscape, it will alter the effectiveness of economics to influence the efficient use of water. Although, as discussed, water is far more than an economic good, balanced allocation between economic sectors, greater efficiency in its usage, judiciously executed infrastructure investment and recognition of the hidden value of water in many globally traded commodities could potentially address some of the social and environmental conflicts of interest raised above.

The views of economists should not be ignored (bad things happen if economists are ignored!). However, it cannot be stressed too much that this is a politically charged area of the economy and, because we do not live in a perfect world, economists can be overruled by politicians.

So, how is the financial credit crunch influencing the economics of water and through it the environmental credit crunch? We can use the same four-part framework (set out above) to track the changes.

Allocation – supply and demand in a credit crunched world

Water consumption basically comes down to three uses: domestic, industrial and agricultural. These demands must compete for water that is supplied by the hydrological cycle and the related infrastructure described above. The financial credit crunch has a bearing on this allocation, as it impacts growth, the desire to continue growing and the demand for food.

The economics

The 'water intensity' of economic activity can be defined as cubic metres of water taken from freshwater sources – measured either per dollar of GDP or per person. If GDP per person rises, and GDP activity also happens to be water intensive, the net result is a rise in freshwater withdrawals per person. As economic activity expands, competition for scarce water resources rises. In 'water poor' locations, that means water usage per head needs to be kept down, either through efficiency measures or by doing less of some things so that other activities can continue. One implication of this is that water 'poverty' could present a constraint on growth in GDP per head.

Most economists would contend that the financial credit crunch will reduce trend rates of growth in the global economy. As such, the pace of increase in water consumption is likely to slow (as water consumption rises with living standards, and slower growth implies a slower rise in living standards). Growth does not, however, cease. The increase in water demand similarly does not cease. This suggests that controlling water demand is likely to be an important part of the solution to ease water constraints.

Household demand

Many in the Western world worry about the quantity of water used in a dishwasher or washing machine, and of course awareness of the consumption of environmental resources is important. However, households comprise 10 per cent of the world's water consumption. Of course, this level has tended to increase with economic development. Household water consumption rises as societies develop (labour-saving devices tend to be more water intensive). For example, African households use 10m^3 of water per person per year (67 bathtubs worth); US households use 100m^3 of water per person per year (or 670 bathtubs, of course).

These bathtubs refer to the *direct* consumption of water by households. They take no account of the water used in producing food and goods that are in turn consumed by a household. We will address those water demands later in this chapter.

Household water demands are likely to increase as economies develop. China's current water consumption is relatively low, for instance. However, as the population becomes increasingly urban, and as standards of living rise, the demand for water is likely to increase. To the extent that growth in the post financial credit crunch world is skewed towards emerging rather than Organisation for Economic Co-operation and Development (OECD) economies (a point we will return to in our chapter on energy), this will mean that every percentage point increase in global GDP is likely to be more (household) water intensive. The aftermath of the financial credit crunch means each percentage point of global GDP growth costs more water, and the environmental credit card will be increasingly used to generate economic growth.

Industrial demand

Industrial use of water accounts for 20 per cent of global demand. Some forms of industry are relatively water intensive (for instance, the manufacture of microchips). Industrial use of water is, like households, in the form of blue water. Industry does not consume water outright – blue water stays blue water through the industrial process, in most instances. However, the form of water is changed by industrial use. Blue water becomes a murky blue, as industry adds pollutants. Nevertheless, because it is blue water that is being consumed, with the right processing water that is used by industry can be reused. The Audi plant in Ingolstadt, Germany, reuses 98 per cent of the water used in car production, for instance.[4] This is an extreme instance, clearly, but it is an instructive reminder of the recyclable nature of water.

The financial credit crunch does mean that economic growth is likely to slow, with all that that implies for the *growth* in industrial water use. The problem is that this is not the only force at work. Politicians, confronted by a more constrained growth profile (and restricted tax revenues at a time of fiscal stringency), are likely to seek to encourage growth wherever possible. Water efficiency is unlikely to be terribly high on the political agenda in this situation. Thus, the financial credit crunch may encourage the ongoing operation of existing water-inefficient industries, in pursuit of economic growth. This is particularly likely to be the case if the *environmental* credit crunch implications fall outside of the national boundaries (foreigners cannot vote, after all; politicians tend to give foreign constituencies little weight). If growth comes at the price of a neighbouring country having less river water, it is a price many politicians will think worth paying.

Agricultural demand

With 10 per cent of water consumption from households, and 20 per cent from industry, the mathematics of water demand is not complex. The big user of water is, clearly, agriculture; 70 per cent of global water demand goes into the production of food. As we demonstrated in Chapter 1, the economics of the financial credit crunch has much to contribute to the debate on food – and through this it has much to say on the consumption of water.

The hamburger has come to symbolize development in many ways. *The Economist* newspaper has its 'Big Mac' index to assess the fair value for international currency rates. Proponents of the 'this time it's different' technological miracle claim that no two countries with a McDonald's restaurant franchise have ever gone to war with each other (Thomas Friedman's *The Lexus and the Olive Tree* published in 1999 put forward this theory). However, the hamburger (whoever sells it) is not just a symbol of development. It is a symbol of water gluttony. A hamburger is nothing more than a condensed form of 35 bathtubs of water – because 35 bathtubs full of water are what it takes to produce *one* basic 100g hamburger in a bun (no cheese, no fries and no 'supersizing').

If societies move towards more intensive meat consumption, then the world will need a lot more water. The newly water-stressed household (sub-31 bathtubs per person per day) will find that it can afford no more than 305 hamburgers in a year, assuming some of the available bathtubs are used for washing, cooking and so on at current US average levels of intensity. Assuming the UN's research is right about the approaching water crunch, there can only be one conclusion: habits will have to change.

Here, the implications of the financial credit crunch can be deemed 'mixed'. If there is a shift in food consumption, in the context of trading down, water may be more efficiently used. However, a key part of this will depend on the ability of economies to trade with one another – a topic covered in the virtual water section.

The economics of water allocation in a financial credit crunch

So, what has the financial credit crunch done for water allocation? It has potentially slowed the growth in water demand (through slower economic growth). It has, however, reduced the efficiency of global economic growth (because of the skew to emerging markets). It is also likely to encourage a more cavalier attitude to water efficiency ('growth at any price', particularly if the environmental price can be passed on to those downstream). Taken together, the financial credit crunch is not likely to increase the (environmental) efficiency of allocating water between the competing demands.

The environment

Current projections suggest that a third of the world's population will not have access to a *natural* source of freshwater in the next few decades.[5] In 2004, 2.6 billion people lacked access to basic hygiene (provided through basic water services). Of that 2.6 billion, 2 billion were from Asia. Consumption of water by households is exclusively the consumption of blue water.

As noted earlier, agriculture is clearly the biggest user of water, and in any allocation of water resources, agriculture has to be the most significant area of concern for the economist or the environmentalist. Agricultural water use is a mix of green water (rainfall) and blue water (irrigation). Around 17 per cent of the world's agricultural output is irrigated. For some specific food sources this level increases further; 40 per cent of the world's grain output depends on irrigation, and thus the management of water.

Food derived from plants (grain, vegetables, fruit and so forth) requires an average of 1200 litres of water (eight bathtubs worth) to produce 2400kcal of food – 2400kcal is the daily calorific content for one person that the Food and

Agriculture Organization (FAO) recommends should be derived from plant sources. However, 600Kcal of meat (again, the FAO's daily recommendation) requires 2400 litres of water (16 bathtubs). Thus, following the FAO guidelines required to achieve a human's requirement of 3000Kcal per day would have agriculture consuming 3600 litres of water, per person, per day (24 bathtubs). Thus – assuming a new technology such as genetic engineering does not rescue the world from this conundrum by delivering a water-efficient seed stock in sufficient quantities[6] the consequences of the financial credit crunch for allocating water resources basically come down to patterns of food consumption.

In fact, agriculture today uses pretty close to 24 bathtubs of water per person per day. The differences start to appear when we break down that figure by region. The differences in consumption patterns follow the standard trend of development. The more 'developed' an economy, the higher the level of consumption of water in agriculture.

If the economic realities of the current climate do lead to a shift in food consumption patterns, this will lead to a shift in water consumption over time, which may increase efficiency of use (averting an environmental credit crunch). However, if political considerations intervene (for example, cheap food, even if water-inefficient or dependent on fossil water) then water is likely to continue to be consumed at an unsustainable pace. This will ultimately mean that the environmental credit crunch could feed back to being an additional constraint on economic activity.

Water infrastructure

If economists are thwarted in their attempts to allocate water resources more effectively across the different sources of demand, can they work on the supply of water? Can improved water infrastructure resolve the imbalances that exist? More importantly, given we are viewing the world through the prism of a global financial credit crunch, does the change in the economic environment have a bearing on water infrastructure?

The economics

Theoretically, a sound water infrastructure in urban cities could make significant differences to the overall efficiency of water provision. In some cities, such as London, some 60 per cent of the water is lost through leaky pipes. This is because the last time a major new build in the water and wastewater infrastructure was undertaken was in the 1800s. Several outbreaks of cholera may have set politicians thinking but what really moved them to action was the so-called Great Stink of 1858, when the waters of the Thames were so foul they enforced a parliamentary recess. Sir Joseph Bazalgette started working on the central London sewerage system the following year.

It is worth considering that the infrastructure of London's water was left largely unchanged for nearly a century and a half. Recently there have been significant programmes to upgrade the existing infrastructure, but the commitment to upgrade infrastructure has been notably lacking for a large swathe of history. Why? Infrastructure costs money, and unless presented with a pressing need (the Great Stink, for instance), there is but slight inducement to spend that money.

Improving the supply of water through infrastructure spending need not cost a lot of money. Rainwater harvesting at a domestic level involves little more than installing a rainwater butt. At around £40 of capital spending, this is not something that is likely to be impacted significantly by the economic climate, at least in developed economies. The problem is that household efficiency in water is the least significant consideration. It is efficiency in agriculture, and to some extent industry, that will make the difference. This is generally more expensive.

Water infrastructure boils down to three basic forms: supply, transmission and end consumption. Supply infrastructure can be primitive (a well is little more than a hole in the ground, after all), or complex (desalination). Transmission is essentially the pipe network, irrigation canals and similar devices. The issue with end consumption is about making industry or households more efficient in their consumption of water (agriculture's efficiency is achieved through better transmission; arguably agriculture could be more water efficient through the species of crops it grows, but this is not really about infrastructure).

Supply of *clean* water also depends on pollution levels. Population centres generate waste, which can raise water pollution levels. As Table 2.1 shows, countries with very large economic output per person put a significant burden of pollutants into water and countries with relatively low economic output per person but very large populations also impose a heavy burden on their water sources. As GDP per head rises, investment in water infrastructure becomes more feasible, and so the 'waste to water intensity' of of each dollar of GDP per person might drop. This is, however, only likely when a certain income threshold has been met. 'Only countries with a relatively high per capita income of at least $2000 per year are in a position to take effective measures for reducing abstractions for public services' (Gleick in Shiklomanov and Rodda, 2003, p31).

Of course, waste in surface water does not reduce the stock of water available. In this sense there is no environmental credit crunch (in the way that irredeemably depleting fossil water is an environmental credit crunch). What is happening here is that the flow of potable water is being reduced.

Water, infrastructure and the environment

Even if the effects of the financial credit crunch go on for some years, the most egregious problems will only shape the landscape for perhaps four or five eco-

TABLE 2.1 Countries with significant emissions to water

Country	*Thousand kgs per day, emissions to water 2005*	*Population, Millions, 2007*	*GDP/Head, PPP 2007*
China	6088.7	1331.4	5380
US	1960.3	303.9	45,590
India	1519.8	1135.6	2750
Russia	1425.9	141.9	14,690
Japan	1133.1	128.3	33,630
Germany	960.3	82.7	34,400
Brazil (1990)	780.4	191.3	9570
Indonesia	731.0	228.1	3710
France	604.7	60.9	33,670
UK	539.7	60.0	35,130

Source: *The Economist* (2009, pp16, 105, 250–254)

nomic cycles (maybe 40 years). Water infrastructure, once complete, tends to be around for many economic cycles. This conflict of cycles means that economics is forced into a subordinate role, and politics is likely to take over. To illustrate this we can consider three rivers. Each is a key part of the ecosystem. Each is also depended upon to support economic activity in several sectors. Development of each, however, is governed by politics.

Shakespeare has his Antony (in *Antony and Cleopatra*) invoke perhaps the most commonly known fact about the river Nile – the importance of its flooding to agriculture: 'As it ebbs, the seedsman/Upon the slime and ooze scatters his grain/And shortly comes to harvest.'

In the 21st century, 96 per cent of Egypt is desert, so, as the population expands (a doubling is projected by the UN by 2050), water is one of the most important challenges. In the modern age, technology allows electricity to be used to move water for irrigation and other purposes, thereby increasing the area of land that can be farmed or lived on in order to accommodate the growing population. This initiative is only indirectly driven by economics – demographics, and therefore politics, is the real driver.

The Amazon is surrounded by a huge area of tropical rainforest, covering some 73 per cent of the river system's 6 million square mile watershed.[7] Some 300,000m^3 per second of water run through the main river at its height, and the Amazon is said to be responsible for one-fifth of freshwater flowing into the sea. Currently, major power infrastructure is under discussion. The impact of such projects on the rainforest, as well as the environment (beyond national borders), is one of several significant unknowns. No doubt economics has come into the discussion on the rationale for this initiative but, once again, politics is not far from the surface.

The third-longest river in the world, the Yangtze, is different again. It is significant as a political boundary (between northern and southern China) in Chinese military history. Historically, it is an important seat of civilization, and its irrigation systems have supported a stable, productive agriculture from the Han dynasty onwards, generating significant economic wealth. It is currently an artery of the transportation system, and also boasts the largest hydroelectric power station in the world, the Three Gorges dam, a relatively recent piece of infrastructure built with economic growth as the main consideration, but with less consideration of its environmental and social impacts. It is possible that these were, to an extent, priced in: as a consequence of private sector reluctance to be involved with unquantifiable (environmental) risk, this project ended up as a World Bank funded project. The real issue driving this project was China's need for sustainable energy – yet again, politics.

The main ways in which an event such as the financial credit crunch can affect such major geo-engineering works is by constraining budgets and altering the cost of capital. This brings us onto the extent to which the financial credit crunch might be relevant to the better usage of water and related infrastructure already at hand: namely, water efficiency.

Incentives for water efficiency

The economics

If water is inefficiently used, can economics provide a remedy for it? Certainly, incentives can be provided that will increase water efficiency. If water is priced to reflect its environmental costs, then there will be a trade-off between the cost of being more efficient (spending on more water-efficient washing machines, for example) and the cost of water consumption. Market forces should then provide the appropriate balance.

Sadly, water is rarely priced to reflect its environmental costs. Indeed, water is generally significantly underpriced. Water rights are also not readily traded. This means that incentives for water efficiency are relatively limited.

For developed economies, household water efficiency is likely to follow the path of energy efficiency. If the consumer pays directly for the water consumed (for example, they have a water meter and do not play a flat rate), then there might be an incentive for them to cut water consumption through small changes. They could, for instance, turn off the tap when brushing their teeth (as the former UK Deputy Prime Minister John Prescott once famously urged them to do).

The problem is that, for developed economies, water is a very small part of the household budget. It is unlikely that consumers would be able to identify the cost of water as an important area for restraint (in the way that they might identify energy costs). Moreover, by limiting the small investments that could be made to improve water efficiency, the credit crunch has, if anything, made attempts to increase water efficiency more problematic.

Box 2.2 The danger of water pricing[8]

Trying to induce water efficiency through market pricing can be dangerous. In 2001, in Kwazulu-Natal, South Africa, the authorities ended the previous practice of providing drinking water for free. Technology was introduced at each communal water tap. Users had to purchase a card, containing a microchip, and 'top up' the card with additional payments, in order to receive water. The principle was much the same as topping up a mobile phone. To get water from a communal water tap, the user had to insert their card, and money would be deducted from the card for as long as the tap was running.

In theory, pricing water in this way should have induced a more responsible attitude to water consumption. People would be aware of the price of water and would have used it more responsibly as a result.

Water efficiency may well have resulted, but the benefits of more efficient water consumption were overshadowed by tragedy. The local population turned to polluted local streams as an alternative water source. Between 2000 and 2002 there was a cholera epidemic and 259 people died.

In 2003, the authorities made the first 6000 litres of water per month free of charge. There is still some slight efficiency incentive (consumers will want to keep their consumption below 6000 litres), but the concept of water as a 'right' rather than a 'need' is emphasized by the free provision of enough water to sustain life.

For emerging markets, the prospect of water efficiency in a constrained income environment is perhaps slightly higher. In some countries (for example, Jamaica, Argentina, El Salvador), more than 10 per cent of household income goes on water charges. However, there are very clear dangers in assuming water efficiency when costs are high, as the Kwazulu-Natal example (see Box 2.2) demonstrates.

Slower income growth (and the attendant desire for thrift in the household budget) may therefore induce some efficiency in household water consumption, but we should not be optimistic. Indeed, developed economies have tended to rely on education and moral persuasion to create water efficiency from households. As for more enduring water efficiency, this seems less likely in the wake of the credit crunch. Purchasing a new washing machine, purely because it is more water efficient (or indeed even more water and more energy efficient) requires a capital outlay. With credit less readily available, and income growth slower, households are going to be less eager to upgrade.

What about efficiency in industry or agriculture? Here, the issue is (once again) the low cost of water, and the relative cost of installing the necessary water-saving devices. What is absent here is the prospect of life-threatening constraints arising from water 'rationed' by price. Farmers may go out of business if their water use is constrained by price setting, but they will not die of thirst.

Agriculture loses vast amounts of water. Between a third and a half of all water consumed by agriculture for irrigation is lost through evaporation or seepage from irrigation canals. Current water pricing regimes, however, leave little incentive to embark on capital investment programmes. The credit crunch and recession further shift the incentive programme. Limited credit and higher borrowing costs mean that major infrastructure changes are unlikely. Governments are also unlikely to promote a more economic pricing of water. Increasing costs (for whatever reason) at a time of slower demand and constrained consumption is not likely to be perceived as sound policy.

The best hope for improved efficiency in water consumption comes from the possibility of a shift in eating habits. If the slower pace of income growth generates more thrifty households, then one possible outcome is reduced meat consumption. 'Trading down' from free-range to factory-farmed meat is also an option. Both of these moves would reduce water consumption by agriculture. The former occurs because non-meat food products are less water intensive. The latter occurs (with other environmental implications) because factory-farmed meat is a great deal less water intensive than free-range meat.

The environment

The usual forces of economics that drive supply and demand (price!) are unlikely to be used by politicians to persuade people to use water more efficiently. This is actually possible within sub-sectors however. The water pricing structure for industrial and domestic sectors tends to be different in OECD countries, reflecting the fact that, as discussed at the start of the chapter, water is a hybrid; social good at one end of the scale and economic good at the other. Whereas it would seem reasonable to use pricing to force industry to use water more efficiently, other incentive structures may be required for agriculture, depending on country-specific conditions, as well as for domestic use.

Anyone who has lived in a dwelling not directly connected to the formal water infrastructure (such as a boat) will already be aware that making water more or less readily available changes usage patterns. So, for households in developed countries, ready availability in the form of a highly developed water infrastructure needs to be paired with an incentive to control usage – such as metering; already a work-in-progress proposal underway in some countries. Moreover, the environment itself (defined as the extent to which the ecosystem provides water resources) can have a powerful effect on the way water is used, forcing more efficient usage. Companies in the aluminium, semiconductor and electricity generating industries in California are competing for limited supplies, and as a consequence have had to embed water efficiency throughout the production process.

As discussed, water usage per head varies hugely from one country to another (America sloshing around in 670 bathtubs of water per capita per year, while Africa paddles in 67 bathtubs). Since water usage developed in the absence of constraints, thereby (albeit unintentionally) encouraging over-use, it is almost certain that water efficiency can be relatively easily improved in the urban environment. Most dwellings in the UK do not recycle the so-called 'grey water' disposed of after washing dishes, doing the laundry or taking a shower, do not harvest rainwater and cannot monitor their water usage.

The problem is that the financial credit crunch reduces the means to 'upgrade' to more water-efficient living, and the pricing mechanism's inducement to pursue such efficiencies is either absent or heavily constrained.

Generally, the financial credit crunch has done little to improve the prospects for water efficiency. Those prospects were never that great in the first place, however, given the uneconomic pricing of water that has prevailed for generations.

Water and trade

When a commodity rises in price, consumers look for substitutes but, for water, once again, conventional economics do not work. There is no substitute for drinking water. Substitutes can be created if water is categorized according to quality. Different amounts of processing could also possibly incur different pricing structures. Indeed, the same idea could apply to different methods of delivery – so, drip irrigation could replace flood irrigation in the production of food, with water priced according to the irrigation method used.

It may be possible to bring about a substitution elsewhere by making sure that the cost of the 'invisible' asset of water in traded goods is recognized. If this were possible, economics would prevail against the arguably profligate use of water in water-poor countries to grow roses for developed-country vases. It could (conversely) encourage the 'export' of water-intensive produce from the water-rich to the water-poor.

The economics

Water is, at least by volume, the most traded commodity in the world. Between 1997 and 2001, 1625 billion cubic metres of water was traded every year on average – a number that rose further in the subsequent decade. This equates to 10,833 billion bathtubs of water. In comparison, the oil industry traded a pitiful 3.2 billion cubic metres of oil around the world in a single year (or 21 billion bathtubs worth).

This is not to say that water is being shipped around in green, torpedo-shaped bottles 'naturally carbonated at source'; a tiny fragment of water is traded through such direct means. Water is packaged quite differently for

Box 2.3 Virtual water

Virtual water is calculated as simply the amount of water that is required to produce a product. Thus, to return to the earlier example, it takes 20 litres of water to produce a hamburger bun (growing corn and so on). It takes between 2500 and 7000 litres of water to produce a 100g hamburger (creating animal feedstuff). Thus, exporting a hamburger de facto involves the export of anything up to 7020 litres of water.[9]

international trade purposes. Water comes encased in grains of corn, embedded into woollen cardigans and as an intrinsic part of the computer microchip. The water that is traded is 'virtual water' – the water that is used as an input into the production of commodities and goods.

The trade in water – as virtual water – is critical to using water more efficiently. As mentioned at the start of this chapter, one of the problems with water is that water in its usable form is unevenly distributed. There is nothing like trade for redressing an uneven distribution (it is why economists get so very excited about the benefits of international trade). If an area experiences water scarcity, then it generally makes sense for that area to import goods or foodstuffs that require a high-water input from other countries; preferably, of course, from countries that have a surplus of water for their own needs. The importing country then does not need to turn to its environmental credit card and run down non-renewable water sources (such as aquifers). The importing country does not need to pursue energy-intensive, expensive forms of water treatment (such as desalination). This avoids spreading the environmental credit crunch from water to energy. We want water (in the form of, say, cucumbers). They have water (and can grow cucumbers). Why not import cucumbers? For some countries, current standards of living are utterly dependent on importing virtual water. Figure 2.1 itemizes (for selected countries) the amount of water consumed (as water and as virtual water) as a percentage of the *renewable* water that country has available.

The extreme is Jordan, which consumes seven times its naturally occurring water supply. Egypt also depends on imported water, and Germany and Spain are intensive water consumers. However, this is just looking at total consumption. If we look at the detail – comparing the domestic and the imported water – in reality, we find that 70 per cent of the UK's water consumption (by the domestic population) actually comes from overseas. The UK has more than enough water to meet its standard of living, but John Bull consumes more virtual water from abroad than domestic water at home.

Virtual water can help the global economy manage its allocation of water more efficiently. With the right sort of trade, water need not be a constraining

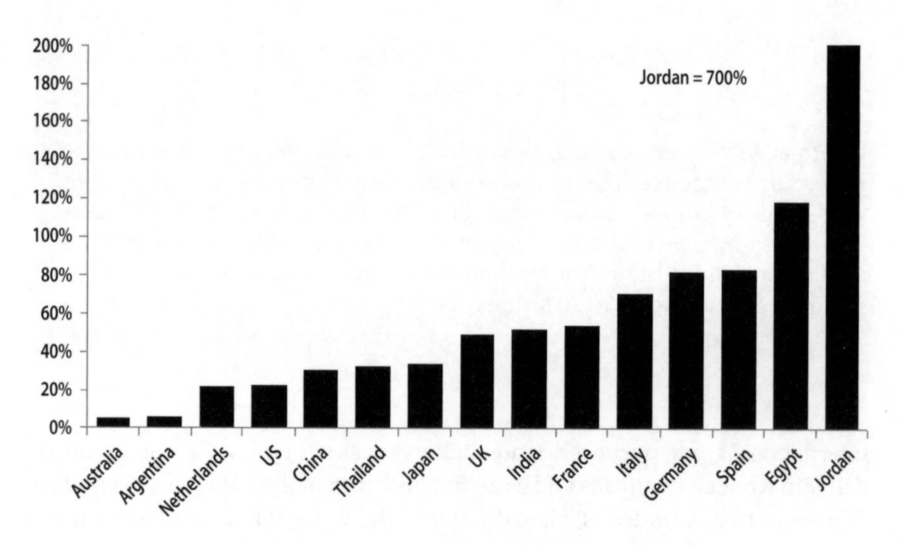

FIGURE 2.1 Water consumed relative to domestic renewable water available

Source: Derived from figures published in Chapagain and Hoekstra (2008, p30).

factor on standards of living. So what does the aftermath of the financial credit crunch mean for virtual water flows around the world? The answer is not good.

The global response to the credit crunch has been to increase trade protectionism. Tariffs have risen. Barriers to trade have been imposed. What is potentially even more worrying is that informal obstacles to trade have sprung up as well. Locally focused banking systems are less able to finance trade – and indeed the whole international trading system effectively came to a shuddering halt at the height of the credit crunch (in early 2009), as the credit that allows trade to take place simply ceased. In 2009, for the first time in decades, global trade declined as a share of the world economy.

As far as agriculture is concerned, this is not a major surprise. Rising food prices in 2008 led to political populism. It became very fashionable to contemplate self-sufficiency in food (autarky) as a desirable policy outcome. Agricultural protectionism is firmly entrenched in the political and wider culture of most countries – think of Japan's Emperor performing rituals to encourage the rice harvest, or the French passion for their *terroir* (the combination of climate, land and position that dictate the quality of agricultural land, particularly in vineyards). Even the British urban middle class cling to a romantic view of the countryside (think of the popularity of the magazine *Country Life* – a means of satiating a lust for the rural life without ever having to leave Islington).

The obstacles to global trade, and the very low starting point for agricultural trade, do not bode well for the prospect of virtual water trade as a means of resolving water imbalances. Some 78 per cent of all virtual water – 8450 billion bathtubs of the stuff – finds its way around the world encapsulated in some kind of agricultural packaging, be it crops or livestock. Although food is the dominant part of virtual water, the prospect for water inefficiency via industrial trade should not be ignored. Indeed, obstacles to international trade in goods pose a serious threat to water efficiency. Fully a third of all industrial water consumption is not for the home market, but for export. This means that the goods that the world exports are actually more water intensive than the goods sold to the home market.[10]

The environment

Any moves to restrict global trade or to continue the restrictions that abound in global agricultural trade, therefore denies the world an efficient allocation of its water resources. Virtual water trade is not the only solution to the threat of water shortages. However, with 16 per cent of the world's water already being traded, it could be a powerful force for redressing the imbalances. The recession and the response of various governments to the credit crunch make that less likely.

Here, then, we see the financial credit crunch exacerbating the environmental credit crunch. Virtual water has huge potential to mitigate a water-specific credit crunch. As we have reiterated throughout this chapter, water is the ultimate recyclable resource – one of the key problems is that its allocation is uneven. Trade in virtual water could overcome that problem. Constraints on international trade (and on trade in agriculture particularly) will prevent any attempts to even out that distribution.

The financial credit crunch tilts the world towards a situation where the environmental credit crunch could accelerate. Countries draw down on fossil water more aggressively in order to sustain their industries or (more likely) their agriculture, for want of anything else. The water sub-category of the environmental credit crunch just gets worse. The environmental credit card bill mounts ever higher.

Water's environmental credit crunch in the wake of the financial credit crunch

The key risk from the financial credit crunch is that political interference in trade and capital pricing will limit measures that could be taken to increase the efficiency of water supply. There are implications for demand for water too – if considerations of economic growth triumph over water efficiency.

Changes in the availability of water will result in its being far more costly (in water terms) to grow certain agricultural produce in some parts of the world than others. This is likely to make virtual water trade increasingly relevant. As the 2008 bans on food exports from a number of Asian countries (a response to concerns about securing food supplies for the local population) have already shown, if this story is framed in a xenophobic fashion, trade may suffer.

Finally, in respect of water, the climate change crunch may turn out to force change. History after all tells many salutary tales of fertile lands turned into deserts, cities rendered uninhabitable and human conflict brought about by changes to freshwater availability.

The 670-bathtub question

The broad question driving this book is whether the financial credit crunch can or will change consumer behaviour to a sufficient extent to move humanity towards an optimal use of this all-important resource. Our work on food strongly suggests that the financial credit crunch will not be enough on its own to bring about a fundamental change to the way resources in general are used. That conclusion matters here, given that agriculture is so important a consumer of water (in this sense, much of the world's demand for water is a derived demand – derived from our demand for food).

As soon as water is used in the context of any economic activity (as opposed to a necessity of life) it becomes a hybrid of social and *economic* wealth, suggesting that pricing mechanisms may be able to come into play somewhere along the line. Its value, from an economic perspective, can be seen as a sum-of-the-parts of universal access (priceless), environmental commons (priceless) and economic good (the economic cost of the effort required to deliver the quality of water needed for drinking or manufacture, as appropriate). This suggests that any attempt to balance the forces of supply and demand will require a hybrid approach; economic forces have a bearing on only one element of the sum-of-the-parts metaphor applied above.

The world is swimming in water, but it is of the wrong sort. The right sort of water is not necessarily in the right place. Agriculture sucks up most of the water. What can be done to ensure the optimal allocation of water between different users? Our assessment suggests that the financial credit crunch is likely to increase the risks of an environmental credit crunch for water. Limits on trade will potentially damage the redistribution of water around the world. Moreover, if the financial credit crunch limits investment, water efficiency could suffer still further. There seems limited near-term hope of floating Earth Overshoot Day towards a later date on a tide of water efficiencies.

The water issue is perhaps one of the most complex we deal with in the list of issues addressed in this book. The issue is not simply one of limited resources requiring rationing. Rather, there are several human behaviour issues: the need to give governments and users incentives to look after the natural water infrastructure properly so as to assure sustainable supplies; the need to share water resources fairly; the need to deal with politics when key water sources are used across regional or country borders; the need to deal with the fact that water cannot be priced properly because it is a 'hybrid' of several quite different social goods; the need to use water efficiently within economic sectors (agriculture, industry, commerce, household); the need to invest in, and maintain, the 'man-made' part of the water infrastructure; and the need to design consumer goods (dwellings and white goods) in order to allow people to make sensible choices about water use.

For now, we must conclude that the financial credit crunch is unlikely to encourage these behavioural shifts. Indeed, to the extent that short-term economic considerations dominate the political process, the likelihood of these behavioural shifts can be thought to be ebbing away.

Notes

1 Data in this section is taken from Mauser (2008, p6).
2 Details of the Mexican situation can be found in *A Guide to the Global Environment: The Urban Environment: World Resources 1996–1997*, a joint publication of the World Resources Institute, United Nations Environment Programme (UNEP), United Nations Development Programme (UNDP) and the World Bank (1996, Oxford University Press, New York and Oxford), http://pdf.wri.org/worldresources1996-97_bw.pdf.
3 The Australian information comes from Peschardt (1999). For more information on the problems of seawater incursion into freshwater river deltas, see 'What lowered water levels do' at http://seawifs.gsfc.nasa.gov/OCEAN_PLANET/HTML/peril_fresh_water.html.
4 The example is given in Mauser (2008, p125).
5 Projections are cited in Larsson (2009, p34).
6 The so-called Green Revolution has brought resource productivity improvement once. The cost of a far more productive industry (in terms of tonnes of food per acre of land) was a heavy increase in water intensity. The 'next' improvement needs to deal with the water problem – but of course any new technology may bring unexpected costs in its wake, just as the Green Revolution did.
7 Data drawn from www.earthtrends.org/maps_spatial/maps_detail_static.php?map_select=410&theme=29.
8 Full details on the Kwazulu-Natal pricing experiment can be found in Gautier (2008, p257).
9 Calculations come from Mauser (2008, p134) and McWilliams (2009, p141).
10 Some 34 per cent of water used by industry is exported. However, total exports (including agricultural exports) amount to 18 per cent to 23 per cent of the world economy, depending on how the measurement is made. Thus, the goods that are exported are using more water than the goods that are consumed at home. Data on water exports comes from Chapagain and Hoekstra (2008). Data on trade comes from the World Trade Organization (WTO).

References

Chapagain, A. and Hoekstra, A. (2008) 'The global component of freshwater demand and supply, *Water International*, vol 33, no 1, pp19–32

The Economist (2009) *Pocket World in Figures, 2010 Edition*, Profile Books Limited, London

Gautier, C. (2008) *Oil, Water and Climate*, Cambridge University Press, Cambridge

Larsson, M. (2009) *Global Energy Transformation*, Macmillan, London

Mauser, W. (2008) *Water Resources: Efficient, Sustainable and Equitable Use* (English translation), Haus Publishing, London

McWilliams, J. E. (2009) *Just Food*, Little, Brown and Company, New York

Peschardt, M. (1999) 'Australia "drowning in salt"', BBC News, 29 June, http://news.bbc.co.uk/1/hi/sci/tech/380907.stm

Shiklomanov, I. and Rodda, J. (eds) (2003) *World Water Resources at the Beginning of the 21st Century*, Cambridge University Press, Cambridge

CHAPTER 3

Powering Down: Energy and the Twin Credit Crunches

> *Day by day it becomes more evident that the [energy] we happily possess in excellent quality and abundance is the mainspring of modern material civilization. As the source of fire, it is at the source at once of mechanical motion and chemical change. Accordingly it is the chief agent in almost every improvement or discovery in the arts which the present age brings forth.* (Jevons, 1866, p1)

Energy runs through all aspects of life, the universe and everything. Without it, there could have been no Big Bang, no life on Earth, and of course there could be no economic activity. Whether directly or indirectly, the physicist, chemist, biologist, climate scientist, farmer, nutritionist and even (unlikely as this may seem) the economist, are engaged in the analysis of their own specialist dimension of energy. Natural stocks and flows of energy are the engine of life; and socially organized energy has been the engine of civilization through the ages as technological innovation has given human beings more efficient means of procuring heat, light and productive work. In essence, organized energy frees up human time and bodily energy for creativity, leisure and consumption.

US President (and Texan) George W. Bush once declared that the US was addicted to oil. In fact, all human beings are energy junkies, even those who are not Texan, and have an apparently unlimited capacity for more.

Of course, human beings were not always energy addicts to the degree of dependence that they have today. Life was described by the 17th-century philosopher Thomas Hobbes as being 'nasty, brutish and short', but life tends to be a lot less nasty, generally less brutish and potentially less short if it is lived in warmth and comfort. As civilizations have developed through history, the human desire for an improved quality of life has tended to lead to an increase in the energy intensity of human activity. Energy is applied in the areas of transport, housing, manufacture of consumer goods and, as described in Chapter 1, in the production of our food. Harnessing energy, almost anything is possible. For millennia, technology-driven step-changes in energy use have also allowed step-changes in human productivity, in turn producing step-changes in economic growth, and in turn producing great improvements in the quality of life.

Keeping up with the Joneses – energy as a limit on the standard of living

The step-changes in energy use only become a problem when things accelerate so rapidly that energy supplies cannot keep up. The current paradigm is that as economies grow, human populations grow; at the same time, energy demands per head expand, and thus human energy needs explode with incredible force. Even a century and a half ago, and before all the energy-consuming gadgets we all depend on were even a twinkle in the inventor's eye, the danger of a relatively fast growth rate in energy needs was apparent to those with foresight. Without coal, said Stanley Jevons, as long ago as 1866, we would be 'thrown back into the laborious poverty of early times'. Jevons was, of course, somewhat ahead of his time in fearing that coal, the new engine of the rapidly growing English economy, could run out and kill off growth at the same time.

It is nonetheless worth remembering that those noble practitioners of the dismal science (economists) follow an important rule of thumb. Predict price or event; or predict timing, but never predict both. Just because a forecast turns out to be off the mark one time round does not mean there is nothing in it. If Jevons were alive now, there is no doubt that he would amend his view of the world as we have done on his behalf in the citation that opens this chapter, substituting the word 'energy' for his original choice of 'coal'.

The power of power

Harnessing energy to deliver heat and light and work to improve the quality of life is nothing new. It goes back millions of years. Campfires help keep people warmer than self-generated body heat could; tractors make growing food easier to accomplish than when done manually; and planes, trains and automobiles make travel and trade easier to do than walking (some of the time). It allows human beings to live in large urban settlements, changing the way work is done and time used, thereby stimulating scientific and cultural development.

Heat and energy

Archaeological evidence indicates that the combustion of biomass (wood, peat, animal dung, and so on) for heating probably goes back to the beginnings of human settlement. Human beings probably started using the fires that are triggered by natural processes such as volcanic activity or thunderstorms well over a million years ago. As Boy Scouts and Girl Guides of more recent times regularly discover, starting a cooking fire without matches can be easier said than done, and even Neolithic man, practised as he would have been in starting fire by friction, may have found it more convenient to keep the home fire burning rather than starting from scratch every time. Matches as we know them today (friction assisted by chemistry) were invented relatively recently, in 1826, by English chemist John Walker.

Light and energy

In a similar vein, the use of waxes and oils for lighting, allowing human beings to ignore diurnal rhythms of dark and light, go back aeons. There is evidence of lamps as far back as 70,000 BC, candlesticks from Egypt and Crete going back to 3000 BC, and also, more recently, tax records referring to tallow candle-makers in Paris in 1292. Candles and early lamps were not very efficient, as anyone who has attempted to read by candlelight in an electrical power cut can testify. Better lighting technology (the use of wick and glass funnel to focus the light) was an important step-change but, of course, unlike lighting technology of the future, still was not sufficient to turn darkness into light for productive purposes. Lace-makers, a century and a half ago, magnified the light of oil lamps by means of water-filled glass flasks, but it is likely that even the most infinitely dim energy-saving light bulbs today generate a better light to work by.

Energy and work

The adventures of Don Quixote – and the famous tilting at windmills story – show us that wind power is a well-established means of harnessing energy for work. Evidence that wind power was quite widely used goes back considerably further than early 17th-century Spain. The Persians used windmills for pumping water and grinding grain in 500–900 AD. In England, the 1086 Domesday inventory of land carried out by the Normans listed 5624 watermills and, by 1300, it is known that there were between 10,000 and 12,000 mills in the UK, driven either by wind or water. In 1750, there were 6000–8000 windmills in Holland and by 1850, there were 9000 of them.

Windmills persisted. The BBC children's television program *Camberwick Green* (first broadcast in 1966) was home to one of the most famous (British) fictional windmills, owned by Windy Miller. Windy Miller was established as the old-school farmer, competing with the somewhat arriviste Farmer Bell (who had battery hens and factory farming techniques). Windy's mill was the most significant visible piece of energy infrastructure in the village, a mere 50 years ago.

Shakespeare's *Henry IV Part 1* (written around 1596–1597) has Hotspur declare (of Mortimer's father) 'O, he is as tedious/as a tired horse, a railing wife;/Worse than a smoky house; I had rather live/with cheese and garlic in a windmill'. Hotspur's somewhat ill-natured character sketch unwittingly provides a perspective on the prevailing energy technology of his day: heat, light and work are a mix of biomass, animal power, human effort and wind power.

Which energy sources went into widespread use in any given country at any given point in economic history had primarily two determinants: what was readily available as a consequence of influences largely out of anyone's control, such as the local endowment of natural resources on the basis of geology, topography and climate; and what could be exploited at the cheapest possible price.[1] That latter characteristic means that energy use falls clearly into the remit of the economist – and it is what makes the financial credit crunch relevant.

History was fuelled organically

In the pre-industrial age, virtually all energy, including the human energy we used to rely on, came from organic sources. After all, humans are simply furnaces for food (albeit operating with varying degrees of efficiency). Thus, the availability of energy was limited by the availability of cultivatable land. The classical economists of this time, living in the context of an 'organic' economy, were cautious when assessing future growth. Unlike the economist of the modern age, they could not predict the scenario of exponential growth that accompanied the arrival of new energy sources (modern-day economists are, of course, almost Delphic in their ability to predict the future). Classical economists could be interpreted as suggesting that the productive capability of the Earth is limited, with the physical limits of land and energy sources being constraints.

Infinite desires and less than infinite resources

As has been mentioned before, economists exist to match potentially infinite desires against constrained resources (it is an economist's sole raison d'être): in the context of energy, this role is concerned with meeting potentially infinite demand for energy by optimally allocating the scarce resources needed for its production. Of course, if the resources are not scarce, no one bothers to do this. In one sense, there is no energy scarcity. The ultimate source of energy for the planet (the sun) is to all intents and purposes an infinite resource.

Unfortunately, this does not mean that energy is an infinite resource. Indeed, this is one of the most toxic parts of the environmental credit crunch – the environmental equivalent of a sub-prime mortgage, perhaps? The *law of energy conservation* in physics states that the total amount of energy in a closed system (such as the universe) remains constant over time; therefore, energy can neither be created nor destroyed. If planet Earth were a 'closed system' (with no energy coming in from the sun), we would not last very long – we may not exist at all. Fortunately for life on Earth, the planet is an 'open' system.

Box 3.1 Peak energy – the crisis that destroyed society

The peak energy crisis, when it came, hit the global economy with an extraordinary abruptness. The consequences were profound. The very structure of society changed. The way people lived, the homes they occupied, the way they ate, travelled, warmed themselves – all were altered in the face of the energy revolution. A form of energy that had seemed abundant, and relatively cheap, was no longer so readily available.

This is not some futuristic projection. This has already happened. The greatest energy crisis is not the 'peak oil' of the future (or the present, depending on which hydrocarbon expert you read). The greatest energy crisis took place more than 90 years ago. The energy that was lost was human labour. The world economy had been geared to the abundance of low-skilled labour for a millennium. With war and epidemic reducing that supply drastically, society was profoundly shaken.

In 1911, 35 per cent of women in the UK were employed as domestic servants. Servants were an absolute necessity in the larger houses of the nobility, the gentry and even the burgeoning middle classes. Cleaning and heating was labour intensive. Food preparation was labour intensive. Further up the supply chain, food provision was labour intensive. The families of dock workers in the slums of East London would find a summer 'holiday' was available if they worked as hop pickers in Kent during the harvest. The harvest needed labour, and families were understandably willing to engage in that labour if it brought with it a release from the surroundings of Cable Street.

In the aftermath of the First World War, the relative supply of labour (as a form of cheap 'energy') dwindled. Servants were not so easy to come by, as the novels of Agatha Christie or Evelyn Waugh's *Brideshead Revisited* testify. It became impossible to carry on living as one had done in the absence of this form of energy – Waugh's fictional Marchmain House in London is sold as impractical by Lord Marchmain. The house is torn down, to be replaced by a 'labour-saving' block of flats. This fictional vignette was repeated in reality across swathes of London. The old form of energy had disappeared.

With the disappearance of one form of energy, a new way of living did produce new substitutes. The transition was painful, disruptive and, at a personal level, often incomprehensible, but it did take place. There is no scullery maid preparing dinner – it comes, prepackaged, stored in a freezer, ready for microwaving. No parlour maid lays the fire each morning in the drawing room – a thermostat automatically adjusts the gas-fired central heating system. Jugs of hot water are not deposited on the bedroom washstand, now that taps readily dispense hot water at will (or at the very least, at the will of the boiler). No groom is required to tend to the horses on a daily basis. A mechanic may service the car annually.

Today, the average American has (in effect) 50 servants.[2] That is to say – the energy consumed by an average American is the equivalent of 50 human beings (to be fair, this includes the energy consumed in food production and other 'higher' supply chain activities). Human energy was replaced with something else: electricity. This drives many of the labour-saving devices that replaced the human hand. Moreover, the energy for heat and light that now comes into every home by far surpasses what could be delivered manually.

Labour saving is not energy saving. It is energy substitution.

Think of planet Earth as a giant battery. When the sun shines on any given surface of the planet, some of the energy that hits the surface of the Earth in the form of heat and light is stored. How much is stored, and how it is stored, depends on what is on the surface. Some will be stored as thermal energy on the land or in the oceans, or as chemical energy in organic matter such as plants, trees and animal life. Some of this energy goes into short-term storage cycles lasting anything from a few hours (the ground warms up and cools down again) to a year or more (plants grow and die back), and these forms of energy are often described as 'renewable' in current parlance.

The 'storage process' for other, mineral, energy forms – the gradual laying down of coal seams and oil reservoirs in rock formations – took much longer and usually took place in extreme conditions that would not easily support life, involving extremely high temperatures and pressures.

The meaning of entropy

From energy storage, we cautiously advance to the concept of 'entropy'. The terms 'high entropy' or 'low entropy' describe energy in a uniquely practical sense. 'Low entropy' means energy is available for use. Examples of substances in a state of low entropy are those that oxidize easily to provide heat or light, such as wood, plant and animal oils, coal, oil, rocket fuel and petrol. When any substance is in a state of 'high entropy', this means energy is no longer available for use. Hence, the ashes left behind once coal has burnt out are in a state of high entropy.

So-called renewable energy sources can (to an extent) reverse the process back from high entropy towards low entropy. So, when the ashes from burnt organic matter are ploughed into agricultural land in order to help produce the vegetation that could be burned for energy, this (partially) reverses entropy. Putting manure on fields to restore the fertility of the soil has the same sort of effect. However, while reversing entropy may be a relatively direct process for organic materials (from plant to manure to soil, and seed to plant), reversing entropy can only be indirect for minerals; ash cannot easily or cheaply be made into coal again.

When *any* raw energy material is used for light, heat or work, it will move from a lower to a higher state of entropy. As discussed, if renewable fuel sources are used, the process can be (either directly or indirectly) reversed; how far reversed depends on the energy source and also to an extent how it is used. If enough energy is drawn down from any particular 'battery' there may be a point at which the 'battery' is not only empty but is no longer rechargeable.

Renewable energy can be replenished (by definition). Fossil fuels are different. They took millions of years to be laid down and, once depleted, cannot be easily replaced. This is essentially what the opening quote from Jevons was talking about.

People have been managing entropy in agriculture since the earliest developments in agricultural technology. As farmers of ancient times discovered, if the same crops are repeatedly sown year after year, the soil gradually becomes less fertile. Crop rotation (first mentioned in Roman times) helps reverse the entropic process in soil as well as bringing other benefits. In modern times, chemical fertilizers are used to the same effect. This is theoretically more efficient than crop rotation by allowing all fields to be constantly in production; but such 'efficiency' may be illusory, first because, even with fertilizers, soil productivity falls over time; and second because the manufacture of chemical fertilizers requires a significant amount of non-renewable resource input, as their name implies.

Using chemical fertilizers, and other mineral sources of energy such as fossil fuels, can be seen as a consumption of past environmental credit savings (and once a savings account has been emptied, it cannot be reused). Of course, if the savings account were very large and the human population relatively modest, this need not be anything to worry about. Unfortunately, the human population is growing rapidly and energy demands per head continue to rise.

Abandoning organic energy

Most of the modern energy that has replaced (or augmented) human energy is in the form of electricity, usually generated by fossil fuels such as coal and oil. The peak (organic) energy crisis caused by a shortage of servants was quickly resolved by an energy revolution. The world just developed gadgets driven by electricity, and switched energy sources.[3]

This drawing down on fossil fuels (effectively consuming the environmental savings account) allowed humanity to break through the restrictions of cultivatable land that had previously hindered energy development. Humanity has no longer had to live within the boundaries set by its environmental 'income'. Alternative energy is the buzzword of the modern age, and it tends to be taken to mean renewable energy. However, in the organic energy era it was the fossil fuel energy sources that were the 'alternative' to the norm.

With fossil fuels and new technology came gains in productivity that economists of yesteryear could only have dreamed of. All production (in whatever sector) in economics requires an energy input in the form of heat or mechanical work. The level of productivity per person that can be reached is strongly conditioned by the availability of energy, whether human, animal or mineral, and so, unsurprisingly, the relationship between energy and real wages and living standards is well established in economics.[4] If energy is cheap, then, basically, human beings can do more, productivity goes up and so do wages. That is the good news for the consumer, at least, as long as cheap energy is around to play with.

No longer cheap or cheerful

The environmental credit crunch, and to some extent the financial credit crunch, changes the price of energy. Cheap energy is no longer assured. What is needed is a change in energy use in two directions: a change in the amount of energy used in economic activity; and a change in the mix of energy sources back to something that looks like the 'old days', from mineral to organic. What we are talking about here is energy efficiency and energy substitution, of course – and these form the very essence of how the financial and environmental credit crunches interact for energy.

The irony for this generation is that history is condemned to repeat itself. Once again, land is the ultimate constraint on energy production and thus the standard of living humanity shares across the planet. The very large simultaneous equation humanity must somehow solve with the help of economics includes food as well as energy: 'Most of the resources for living sustainably are related to land area: if you want to use solar panels, you need land to put them on; if you want to grow crops, you need land again' (MacKay, 2009, p117). The need is therefore to use resources as efficiently as possible, as we stop running down unsustainable forms of energy credit in order to live within our environmental (and economic) spending limits.

Energy and the twin credit crunches

The energy crisis is the leading example of the environmental credit crunch. Could the financial credit crunch bring about a belt-tightening in energy use? Can the full arsenal of weapons manipulated by professional economists (the forces of supply and demand, basically) bring about a structural break in the consumers' hedonistic spending of environmental credit?

The financial credit crunch has implications in three key areas: for energy efficiency; energy substitution; and above all for energy infrastructure (which feeds back into the first two implications). It is time to think about the financial credit crunch and how the economics are likely to work in the context of energy.

Energy efficiency after the financial credit crunch

The economics

The big picture for economics seems simple enough: the world economy is growing more slowly. Indeed, in 2009 (for the first time in decades), the global economy actually contracted. If there is less economic activity taking place, there will be less demand for energy in order to power that economic activity. Moreover, if the trend rate of growth in the world economy is lowered as a result of the financial credit crunch, the slower growth of the world economy will lead to a slower rate of growth in demand for energy in the future. That eases the strain of the environmental credit crunch (there *is* still a strain, but it is less acute).

So far, this is all fairly clear. Of course, it is worth underscoring that the demand for energy is not likely to drop in the years ahead. The world economy will grow at a positive rate (it is generally to be hoped), which means that energy demand growth is also likely to be positive. Nevertheless, there is a slowing in the growth of energy demand, which means we should not extrapolate from past trends.

Unfortunately, the big-picture consequences of the financial credit crunch are more complex than this overview suggests. All economies have slowed, but some economies have slowed proportionately more than others. Generally speaking, it is the wealthier economies of the Organisation for Economic Co-operation and Development (OECD) that have suffered more in this economic crisis. For the first time in a long time, emerging markets have not underperformed (economically) as the world economy has slowed. Normally, we would expect the US to sneeze, and the emerging markets of the world to be stricken with a case of double pneumonia. Not so this time. The US and the rest of the world seem to have the same virus. The margin of difference between more rapid emerging market growth rates and OECD growth rates has not narrowed in the way that it has done in the past.

Why does this matter? It all comes down to energy efficiency. Basically, it takes more oil to produce a dollar of gross domestic product (GDP) in China than it does to produce a dollar of GDP in Germany. China is less efficient in turning its energy inputs into economic output. Therefore, the composition of global growth is going to matter a lot to the macro-cyclical pace of energy demand. The structure of the downturn to date, and the likely impact on trend growth, suggests that emerging markets will outperform OECD economies in their contribution to global growth.

The financial credit crunch therefore suggests that the growth in global energy demand will slow, but not as much as the pace of growth in global GDP will slow. (Of course, this assumes that current relative trends between emerging and developed economies' growth rates continue. They might not. More trade protectionism policies or less foreign direct investment could yet infect emerging markets with the economic equivalent of pneumonia.) The emerging pattern of growth, with its bias to emerging markets, suggests that each percentage point of global growth will be made up of less energy-efficient components than has been the case in the past.

The environment

The impacts of energy efficiency and a slower macro economy on the environmental credit crunch are pretty obvious. US electricity demand failed to grow at all in 1974, in the face of negative growth and a weak consumer. The collapse of communism led to a 37 per cent reduction in Eastern Europe's greenhouse gas (GHG) emissions between 1990 and 2006 – because it led to a collapse in the economies of Eastern Europe (led by a collapse in the consumer sector). We know that demand for energy is a demand that is derived from economic output. We also know that this is not necessarily a desirable means of achieving lower energy demand.

In the 21st century, energy efficiency is viewed as the cheapest available source of alternative energy. Energy efficiency means using energy as frugally as is possible *at each stage of the energy food chain* – generation, transmission and use. The force to be more frugal ultimately resides with the final link in the chain – with 'use' (the consumer, in other words) – which is why the consumer's reaction to the financial credit crunch becomes so pertinent.

For energy generation, this means using fuel-burning techniques that extract the maximum joules per unit of fuel, and that do not waste heat by sending some of it up the chimney into the air along with waste gases. (A gas-fired power plant built in the mid 1990s is 35 per cent efficient. A modern gas-fired power plant is 60 per cent efficient – that is to say, 60 per cent of the energy in the gas is converted into electricity.)[5]

For transmission, this means designing the wires that take electricity to where it is needed so that one watt sent down the wires is as close as possible to one watt at the point of delivery. (HVDC – High Voltage Direct Current – is a technology that reduces power lost in transmission over long distances.)

For usage, energy efficiency simply means using the minimum amount of energy per mile travelled, per home heated, per units of industrial output or per tonne of agricultural produce delivered. If consumers are sensitive to the economic cost of energy (because of the financial credit crunch), they will demand the most efficient use of energy possible in their day-to-day lives. That, of course, means efficiency in usage, but also forces efficiency on the earlier links in the chain.

Waste not, want not – waste and energy efficiency

We saw with food that one consequence of the financial credit crunch may be a reduction in waste as consumers become more sensitive to the cost of waste. This matters in energy consumption. If waste is reduced, then energy is used more efficiently. Economists would say that increased efficiency means that less energy is used to generate a unit of economic utility (economists say these things; it is best to let them get it out of their systems). Putting it another way, there are fewer joules consumed per buck spent.

'Waste' can refer to material that is potentially economically useful that is not used or it can refer to so-called 'waste products' – material that has no (perceived) economic use and is therefore abandoned – emitted to land, air or water. The familiar saying from the north of England runs: 'Where there's muck, there's brass.' Something one person has no use for and throws away, might be very useful to someone else. There is an increasing awareness of this point in the 21st century.

The awareness of waste and its value is a developing concept. Charles Dickens' *Hard Times* (first published in 1854) is set in the fictional locale of Coketown 'a town of red brick, or of brick that would have been red if the smoke and ashes had allowed it'. However, the consumers of the produce of Coketown deliberately sought ignorance as to the environmental waste of the place – a typical customer was a 'fine lady, who could scarcely bear to hear the place mentioned'. Today's consumer – aided by television and the internet – is more aware of the environmental waste that their consumption can produce.

One example of awareness of the value of waste is in the recovery of rare metals for recycling from diffuse sources. Platinum thrown from the exhaust pipes of cars, rare metals embedded in abandoned mobile phones; these are increasingly likely to be a gold mine (in a very literal sense in the case of electronic products) for whoever discovers an economically viable means of recovering these metals. If not recovered, they can be regarded as *energy* thrown down the gutter, whether from the perspective of embedded energy (entropy), or from the perspective of economics: the energy used to process, extract and install these metals in the relevant gadgets is thrown away if they are not recovered.

Restrictions on waste disposal puts up the costs and may help tip the balance towards a more efficient use of energy (and, indeed other environmental resources). In a financial credit crunch, increasing consumer sensitivity to expenditure means that consumers are likely to be increasingly aware of the costs associated with waste. It is not that the environmental credit crunch has become any more acute – merely that the financial credit crunch makes the environmental credit crunch more visible to the average consumer.

Carbon, costs and efficiencies in the wake of the financial credit crunch

In a similar manner, existing (economic) costs associated with carbon emissions may encourage efficiency, as the financial credit crunch heightens consumer sensitivity to any economic costs.

It is known that several countries have 200 years or so of coal stocks and, even if oil production has hit a peak, it will be many years before we have to worry about running out. In theory, this gives plenty of time to find ways of replacing fossil fuels with something else.

However, existing regulation means that the cost of emitting greenhouse gases is relatively high. It is also noticeable that the financial credit crunch, for all its economic horrors, has not induced a reversal of these regulations to date. In the next five to ten years it is likely to become very expensive to emit greenhouse gases, and the consumer will react to that increase in price with a desire for greater efficiency.

It is easy to dismiss regulation as something people simply try to get around, but it may actually be very effective. The UK Clean Air Act of 1968 (which consolidated an earlier act of 1956) happens to be a good example, clearing the air of particulate matter injected into the air by coal burning. One of the authors is old enough to remember a striking impact of the Clean Air Act. In the 1960s many of the buildings in northern Lancashire towns were black. As a child, the author believed this was the natural state of affairs, until some of the buildings were cleaned post 1968, stripping off layers of dirt deposited during the earlier years of coal burning (without emissions controls), to reveal as if by magic the stunningly beautiful golden stone underneath.

Pollution in this instance was a very visible form of waste and inefficiency. Coupled with the price increase for pollution (be those carbon emissions or other forms), consumers have an economic motive to demand energy efficiency in the products that they buy. This means they are either demanding energy-efficient products per se (with a lower running cost), or alternatively products that have themselves been manufactured in a more energy-efficient manner (and which thus have a lower price).

A better way to be efficient – substitution and its prospects

The economics

The substitution effect on energy demand is more subtle than the rather brutal consequences of trying to reduce consumer demand (and thus energy demand) outright. Essentially, the substitution process is all about changing consumption patterns, rather than an absolute reduction in demand. Consumers, in particular, seek to use cheaper forms of energy, or to use their energy more efficiently, to achieve the same standard of living.

How does this work? Imagine an economist, sitting at home in the winter. With high fuel oil costs, no prospect of a pay rise and concerns about the economic outlook, there is a strong incentive to cut expenses in the home. How can this be achieved? Expenses can be cut by turning the thermostat down by one degree and putting on a woollen sweater. Reducing the thermostat a degree may reduce the home heating bill by 10 per cent. The home is colder, but the economist accepts that (and uses their personal body heat more efficiently, by effectively lagging that particular form of heating through the wearing of a sweater).

The economist may also become conditioned to such efficiencies. An economist in the time of Keynes would have been shocked at a living room heated to 21°C in the winter months. An economist in the 1930s, crouching over a coal fire in their lonely garret, would not be able to imagine such a temperature as being not only normal, but recommended. A suit from the pre-war era has insulating properties that are entirely absent from most of today's clothing (and a tweed suit from the 1930s seems thick enough to serve as body armour today). Turning the thermostat down a degree today is a minor alteration against such a benchmark. The consistent warmth of housing in OECD economies today is unusual.

The consumer's incentive to cut energy costs is relatively high, because energy is (like food) a high-frequency purchase. The consumer is therefore disproportionately sensitive to energy prices. The most obvious sensitivity is to petrol prices, which is a very high-frequency purchase for car-owners. Even non-motorists cannot generally escape sensitivity to petrol prices, because they are displayed with such prominence. This sensitivity to the price encourages a desire for energy economy in an economic downturn.

Of course, it is not just the central heating thermostat that is a target. Consumers may be tempted to lower the temperature of their washing machine (encouraged by soap manufacturers' advertising claims) or to switch to lower-energy bulbs. Consumers may also seek to find substitutes for the transport energy represented by driving a car. In 2008, local media on both sides of the Atlantic focused on schoolchildren riding horses to school in order to save fuel costs. This is not all necessarily energy *saving*; some of this is a form of energy substitution. Horse fodder is a form of energy, and the production of horse fodder may even involve oil as an input.

The substitution effect is a way of reducing energy demand. The problem is that it may not endure. In February 1977, President Carter donned a beige cardigan and appeared before the US nation to urge that thermostats be reduced, as a response to the 'permanent' energy shortage that the nation faced. Despite a positive reaction to this populist appeal, the US consumer still consumes 25 per cent of the world's energy (with around 5 per cent of the world's population). Carter's cardigans did not inspire that much efficiency. Consumers can become inured to higher energy costs. Most people live without 'energy poverty' (which occurs if 10 per cent or more of a household budget is spent on energy). For such households, energy costs are not too dominant a form of expense. Turning the thermostat up a degree for an evening seems to be a relatively small price to pay (particularly if the alternative is getting up off the sofa to walk all the way upstairs to fetch a sweater – a feat of athleticism that is beyond the fitness capabilities of just about every teenager in the country).

Although energy is a high-frequency purchase, in the form of fuel for the car, the frequency and visibility of energy costs for most consumers is less. Monthly direct debits at a fixed rate mean that the association of today's central heating warmth with tomorrow's gas or electricity bill is less than direct. The old 'meter' approach for payment (or its modern equivalent for lower-income families, the 'key meter') gave a more direct association between consumption and cost. So, unfortunately, it looks as if the credit crunch will not help the planet much.

The environment

Economic forces led to a substitution of fossil fuel sources of energy for renewable energy in the Industrial Revolution. When this happened, the environmental externalities associated with the widespread burning of fossil fuels were not understood and were certainly not captured in the economic price (and thus had no economic impact). The challenge for the new millennium is to reverse the process, substituting high-carbon fuel sources for low-carbon fuel sources. Specifically, the question is how to leverage the forces of economics to render alternative energy a viable resource, given economic needs.

Once again, the problem for substitution is to address several simultaneous constraints (such as the availability of land and raw materials). Undoubtedly, the consumer, as the ultimate arbiter of economic activity, will have to play a central role in the solution.

In the pre-industrial age, land was the main restraint in the context of energy. Land remains the main constraint in the modern age without the escape route of coal and oil this time around. However, since the pre-industrial age, new technological developments have made it possible to leverage energy sources in ways that could not even be envisaged in pre-industrial times. While the limits of technology to solve resource constraints must be recognized, the fact is both wind and solar power can be leveraged far more effectively than they have ever been. The received wisdom is that wind and solar energy can never be more than marginal, but it is unlikely that the limits of these energy sources have been fully explored.

The most salient example is so-called concentrated solar power (CSP), which is capable of delivering 15 watts per square metre, far in excess of other sources and therefore the most 'land efficient'. Yet, as we write, relatively few CSP plants are installed. The Mojave Desert plant in the US has been producing electricity since the mid-1980s, so the question is why are there not a lot more such plants?

The price of energy substitution does not yet reflect the true environmental costs of failing to substitute. The environmental credit crunch is not 'in the price' that economic (market) forces put on energy. This means that those economic forces for energy substitutions that *do* increase in the aftermath of the financial credit crunch are blunted. The economic forces from the financial crisis act on the economic costs for energy. If the economic costs were greater (as they would arguably be if they incorporated environmental costs), then the power of economic forces to bring about (environmentally driven) energy substitutions would be correspondingly magnified.

However, there is an additional consequence from the financial credit crunch for the environment. This helps to explain why the level of substitution is so low to date – why the Mojave Desert is an exception and not the rule. Perhaps the most serious consequence of the financial credit crunch is the impact on society's ability to effect energy substitution via the impact on energy infrastructure.

Infrastructure and energy in the wake of a financial credit crunch

The economics

The impact of the financial credit crunch to date has focused on the way in which the economic cycle impacts energy demand. While energy efficiency takes us some way beyond pure cyclical effects, the impact of the financial credit crunch in this area is still largely cyclical. However, potentially the most serious impact of the financial credit crunch on energy demand is the ability to undertake large-scale structural changes.

Let us return to our economist, reducing the central heating thermostat by a degree. This reduces the central heating bill, to be sure, but there are even more ways that energy efficiency could be achieved. The economist could purchase a modern boiler (most boilers more than ten years old are significantly more inefficient than their modern equivalents). Indeed, the economist could substitute an oil-fired boiler for a woodchip boiler – increasing efficiency (with a more modern boiler) and switching from a non-renewable to a renewable energy source. Indeed, the economist could go even further and install wood-burning stoves in all the rooms in the house, thus heating only the parts of the house that required heating – and again substituting non-renewable energy with renewable energy.

These measures represent step-changes in energy consumption. However, the reader will note that this is a different step-change. In the past, step-changes raised the standard of living – a stride forward, as it were. Such step-changes as these are more akin to running on the spot – the change preserves the existing standard of living.

All of these step-changes are excellent suggestions. All of them will save the economist money. All of them reduce the effect of the environmental credit crunch – and push out Earth Overshoot Day to some further point in the future. And all of them will require an initial financial outlay. (However, the discerning environmental reader will have noted that if everyone followed the lead of the sweater-clad economist abovementioned and burnt wood for all heating needs, we might run into a wood crunch just as the Victorians did, unless a sustainable source of biomass could be identified.)

While the energy efficiency may mean that the investment 'will pay for itself in ten years', this is not much help if there is no means to raise the initial sum required.

The cost of borrowing money

The most potent impact of the financial credit crunch is an increase in the cost of borrowing money. Lenders have reassessed risk and, relative to the ease with which credit could be obtained over the past decade, money is now hard to come by. This means two things. First, it is more difficult to borrow a sum of money at all. Second, the cost of borrowing that money today is higher than it would have been for an equivalently situated person in the past.

This means that any increase in the cost of energy (either in absolute terms, perceived terms or relative to household income) meets the absolute increased cost of financing the capital investment required to become meaningfully more energy efficient. The desire to become structurally more energy efficient may increase, but the cost of doing so rises. Indeed, if the financial credit crunch is particularly severe, then the ability to borrow money to make the necessary capital investment may be denied entirely.

Of course, one could simply save the money to make the investment – but here we enter a Catch-22 dilemma. If one has to save up to buy a wood-burning stove (or whatever piece of capital equipment is desired), one has to cut spending elsewhere. If, however, the cost of energy consumption cannot decline until one has made the new capital investment, it is more difficult to save, and so one is less likely to make the energy-efficient capital investment at all.

The dilemma of the shivering economist at home, or the motorist driving their petrol-guzzling car, is magnified when we consider the national need to find alternative forms of energy and to use them efficiently. Consider the costs of running a nuclear power station in the UK. On average, roughly 10 per cent (depending on precise conditions) of the costs are fuel – which is no small consideration, of course, though a great deal less than for a coal or gas-fired power station. Estimates of back-end costs such as waste storage can be as low as 3 per cent (depending on the assumptions used to calculate it), and a further 20 per cent represent operations and maintenance. If roughly one-third of the costs of running a nuclear power station are represented by the actual day-to-day requirements of running it, this means the lion's share of the running costs for a nuclear power station come from capital costs – the funding of the infrastructure in the first place. In terms of its cost profile, the nuclear power station is the national equivalent of the oil-fired boiler in the home.

With so massive a capital outlay involved, a higher general cost of capital in the world economy makes investment in nuclear power an unlikely substitute for the status quo. The problem, of course, extends beyond nuclear. Consider the gas generation of electricity, introduced earlier: gas power stations today are nearly twice as efficient as the gas power stations of the 20th century. So dramatic an efficiency improvement should suggest that, at the very least, older gas generators are scrapped or upgraded. The capital cost involved has become a larger obstacle to that attempt to increase efficiency. This impacts the consumer because, of course, the consumer is denied cheaper energy (cheaper in economic *and* environmental terms) by constraints on the government's ability to finance infrastructure spending.

The environment

Economics has few tools for dealing with the indirect effect of existing (non-energy related) infrastructure on the way energy is generated, delivered and used. Transport, the built environment, the manufacturing capital stock and the design of thousands of consumer products were put in place in the presence of abundant energy supplies and on the assumption that cheap energy would continue to be available for the foreseeable future. Even if individuals want to change the way they use energy, what they can do may be limited. The commute to work, to the supermarket, the school, is hard to avoid. Those seeking to get out of the car and onto the train, soon discover that the time constraints in their daily lives (the working patterns typical of the modern economy) do not give them the flexibility to use public transport easily. Train timetables often appear to belong to a gentle era in which the unhappy economist did not get up at the crack of dawn to arrive in the office before trains begin running.

Of course, not all consumers are so constrained, but, in general, given the choice, the travelling consumer shows a preference for the car: it is often less expensive than other forms of transport, it is more comfortable and it is obviously more flexible. A significant investment in public transport infrastructure is required, in some (but not all) countries before it becomes a meaningful substitute for the car. Consumer tastes can change of course, so consumer habits are more flexible, but even they can be considered structural if they are well-entrenched.

Investment in more-efficient or more-renewable forms of energy today simply requires existing technology to be used. Future gains for the global economy, however, require more-efficient technology to be discovered in the future. Economic growth today must be more innovative (doing more with the resources we have to generate more output, not increasing the resource inputs to generate more output).

Even in 2006/2007, when global credit was still flowing freely, research and development spending in electricity amounted to 0.9 per cent of sales. With profits challenged and finance less forthcoming, that figure has to be considered vulnerable. Yet, even the limited spending on research in the past was capable of generating some significant efficiency results. The International Energy Agency estimates that energy efficiency improved 1 per cent every year in the recent past. From an environmental perspective, that was hugely important – 36–44 per cent of improvements in greenhouse gas emissions are attributed to greater efficiency in energy, rather than to shifts in energy production.

Investment in energy efficiency is therefore important, has the potential to yield significant economic and environmental gains, and yet is challenged (as is any form of investment) by the financial credit crunch. There is a further compounding problem, in the form of structural impediments to structural change, and unfortunately the credit crunch is likely to keep such structural impediments firmly in place.

The financial credit crunch, the environmental credit crunch, and energy

That 70s show

The most significant energy crunch in living memory for current generations took place in the 1970s oil shock, and this brought about a significant change in behaviour. We must face reality – the environmental shifts in behaviour were triggered by a price shock. Money talks. Price shocks are extremely effective in bringing about changes in human behaviour. This is one reason the financial credit crunch of the past few years could turn out to be good news for the planet – it is an income shock, which increases sensitivity to price.

Many consumers today are too young to remember the seismic impact on daily lives that came out of the oil crisis of the 1970s or the coal strikes in the same decade in the UK. Most British consumers will remember the long lines at petrol stations in 2008 when petrol delivery drivers went on strike. Anyone dependent on gas will remember the Ukrainian problem at the start of 2009, which put European gas supplies from Russia at risk. These serve to remind us of what life might be like without reliable energy supplies. This is the consequence of maintaining today's standard of living on the equivalent of an environmental credit card.

Throughout history, the energy source relied on by society was whatever was easily and cheaply available. The same narrow economic considerations have driven developments in energy technology. History is scattered with examples of the decline and fall of civilizations brought about by resource constraints and environmental degradation. The coupling of energy with rising material prosperity observed in most economies as they shift from emerging economies to developed economic status encapsulates the problem in focus in this book.

Cautious pessimism

In essence, history also shows that as soon as industrial revolution arrives, policymakers ignore environmental issues in the drive to leverage cheap, abundant energy to drive economic growth. Something needs to put the brakes on this process. Given the need for a braking mechanism, the question is whether the protracted 'credit corset' many of us are about to experience, and indeed are only in the early stages of, could be at all significant, as a braking mechanism, from an environmental perspective.

Will the slower growth in the levels of material prosperity that follow from the financial credit crunch put less strain on limited energy resources? The cyclical consequence of slower growth in the future is slower growth in energy demand. However, energy demand is not *reversed*. The micro-level changes (turning the thermostat down a degree) may endure – even if the lessons of history are less encouraging. Even if they do endure, however, the leaps of energy efficiency we could make with existing technology become less likely as capital becomes less available.

Will behavioural change triggered by the financial credit crunch result in less energy waste: fewer old cars sent to scrap, smaller old tyre mountains, less electrical waste sent off to China for de-manufacture? This is possible – waste is an obvious enemy in a new-found age of austerity. However, targeting waste alone is not enough to bring an end to the environmental energy credit crunch.

Will the financial credit crunch lead to a rethink of economics? Could there be a paradigm shift away from the narrow productivity obsession of politicians (and some economists) towards something like 'green' (less energy intensive) GDP? This seems unlikely, but some aspects of 'green' GDP may creep into traditional GDP numbers through formal environmental rationing schemes.

More significant still, the unknown potential improvements in future energy efficiency are rendered still more distant prospects as research and development falls victim to reduced capital spending. However, innovation can (even if credit constrained) still accomplish change if the pricing mechanism encourages it. The great energy crisis of the 1920s led to innovation and social revolution. 'Labour-saving' devices were the objective of the age, because domestic labour (as the form of energy for ages past) is no longer readily available at a cheap rate. The shift in relative prices led to innovation and substitution, even through the constraints of the 1930s. Such innovation could come again.

There are two grounds for optimism. Policy, properly implemented, may yet turn this around. Great strides in energy efficiency came out of Europe in the wake of the oil crises of the 1970s. Huge economic changes resulted in a very short space of time during the Second World War. Governments can accomplish much with policy, and they can accomplish much in a short space of time, if the incentive is strong enough. Balancing energy supply and demand in sustainable ways is above all an issue of social organization. It is civilization that is at stake.

What is certain is that change must happen. Remember the US citizen and their average 50 servants (disguised as non-renewable fuel)? Let us suppose that renewable energy meets 10 per cent of US demand in 2020 (an ambitious hope). The average US citizen would then have renewable energy providing the equivalent of five servants. They would have their own energy as well. That would result in the energy equivalent of six people. The energy equivalent of 44 servants would still be consumed by the US consumer in an unsustainable manner. Losing the efforts of 44 servants would represent an awfully abrupt adjustment to the hapless consumer's standard of living.

Notes

1. This could lead to innovation. Cornish tin mines required efficient pumps because of their susceptibility to flooding. However, the absence of coal in Cornwall meant that it was expensive to operate steam-powered pumps. As the energy required to pump deeper mines made steam power the only viable option (given prevailing technology), there was motive for seeking greater efficiency from the pumps.
2. See Larsson (2009, p60), quoting Richard Heinberg's calculation.
3. The question is whether, in the absence of coal and oil, human ingenuity might have found other (organic) means of procuring energy. Almost certainly such a discovery would have taken a lot longer to uncover. Technologies such as concentrated solar power were certainly not beyond the scientific knowledge of the Industrial Revolution, but the world had not shrunk to the extent it has today, and high-voltage transmission grids bringing energy from hot deserts were, and for that matter still are, an idea of the future. It is perhaps worth noting that the 18th century saw water mills increasing their efficiency at the same time as steam engines combined heat and work for the first time. Ingenuity led to organic energy improvements, but these gains were eventually trumped by inorganic energy. See Mokyr (2009, p127).
4. See, for example, Wrigley (1988). One economic historian, tellingly, comments: 'The ecological roots of the Industrial Revolution are not difficult to find. The initial stimulus to change came directly from resource [wood] shortages ... due to an economic system expanding to meet the needs of a population growing within a limited area' (Allen, 2009, p80, citing Wilkinson, 1973, p112). The industrial age can be seen as having been facilitated by the arrival of coal. At the same time, however, the arrival of coal was actually a solution to a problem of environmental constraints on economic growth.
5. This data is cited in Woudhuysen and Kaplinsky (2009).

References

Allen, R. (2009) *The British Industrial Revolution in a Global Perspective*, Cambridge University Press, Cambridge

Jevons, W. S. (1866) *The Coal Question: An Inquiry Concerning the Progress of the Nation, and the Probable Exhaustion of Our Coal-Mines*, Dodo Press, Gloucester

Larsson, M. (2009) *Global Energy Transformation*, Macmillan, London

MacKay, D. J. (2009) *Sustainable Energy – Without the Hot Air*, UIT, Cambridge

Mokyr, J. (2009) *The Enlightened Economy*, Yale University Press, New Haven, CT

Wilkinson, R. (1973) *Poverty and Progress: An Ecological Model of Economic Development*, Methuen, London

Woudhuysen, J. and Kaplinsky, J. (2009) *Energise*, Beautiful Books Ltd, London

Wrigley, A. (1988) *Continuity Chance and Change: The Character of the Industrial Revolution in England*, Cambridge University Press, Cambridge

CHAPTER 4

Infrastructure: Building the Future in Credit-constrained Times

When the capital development of a country becomes the by-product of the activities of a casino, the job is likely to be ill-done. (Keynes, 1936)

If an economist were forced to narrow down the consequences of the credit crisis into the single most salient point, the conclusion would almost certainly have to be 'investment in infrastructure'. Infrastructure investment is the building and maintenance of the basic organizational structures needed for most of the activities of everyday life. Any activity that requires energy, water, storage, transport or physical support will require economic infrastructure. Demand for infrastructure is, of course, a derived demand. No one wants infrastructure for what it is. Infrastructure is what supports the modern economy – and thus it depends on consumers' demand for other products. Consumers dictate what the infrastructure of an economy needs to provide, and the availability of infrastructure will likewise shape what consumers can consume – which is why infrastructure is pertinent to this book.

More than anything else, the financial credit crisis has threatened to restrict the level of investment in infrastructure that can be expected in future years – and at just the point when population growth, changes to the demographic mix, urbanization and the environmental impacts that come in their wake demand *increased* investment. The investment consequences of the financial credit crunch were never going to be helpful, but they have come at probably the worst possible moment in modern economic history – for it coincides with the onset of the environmental credit crunch.

Any solution to the environmental credit crunch that maintains humanity's standard of living *requires* an 'upgrade' of the economic infrastructure.

What defines infrastructure?

It is worth starting with the observation that infrastructure is not just the built environment we see around us. Environmental infrastructure is taken for granted, but surrounds us. Life without the built infrastructure we unthinkingly use every day would be a lot more uncomfortable; human life without the natural infrastructure of the planet would, very simply, not be here at all.

When there is no life on any given planet, its infrastructures – defined as the basic organizational structures that contain or shape it – tend to be relatively simple. As soon as life-supporting substances appear, such as air and water, things become far more complicated. Water, rock and air, and the life forms they support, interact to create a new infrastructure that, over time, becomes more and more organic and, as a consequence, increasingly complicated. Man-made infrastructures often mimic natural infrastructures; reservoirs mimic lakes for instance.

Nature's infrastructure serves several basic purposes: support, containment, transport and transmission. Infrastructures made by life forms such as ants, birds, moles, wasps and, of course, human beings, whether by accident or design, mimic the shapes that perform these functions within the ecosystem. If this sounds improbable: the rivers and lakes would still be there to contain water; chimneys, channels and lakes would still form to contain volcanic effluvia; and natural balconies (technically speaking known as travertines) would have formed on the mountainside overlooking a spectacular view in Pamukkale (Turkey), with or without the presence of human beings to appreciate their beauty or to mimic them in building the structures needed for support, containment, transport and transmission in everyday life.

The infrastructure that is relied upon by human beings to support their daily existence does not have to be man-made. As discussed in the earlier chapter on water (Chapter 2), the great cities of the world tend to develop in the presence of some of the most important pieces of natural infrastructure: rivers, lakes and seas. Such cities can change shape or even disappear if there are permanent changes to the natural infrastructure (as was the case with China's Black City).

As soon as life forms appear on the planet they have an influence upon its infrastructure, shaping and expanding aspects of it, consciously or unconsciously, to accommodate them. Some forms of influence are more intrusive than others (as the case study in Box 4.1 demonstrates).

The relationship between life on the planet and the planetary infrastructure could be described as symbiotic, except for the point that when organisms live in close association with each other it is often to the benefit of both, whereas most of the unplanned human impacts on the environment tend to be negative for the environment. Even if it is stretching it a point too far for some to see the entire planet as an organism, it is stating the obvious to say that human beings need the planet's natural infrastructure, but the natural infrastructure does not need human beings.

Nevertheless human beings should take note of the point that there does appear to be a complex multi-directional relationship between the living planet and some of its other life forms. In his seminal writings on the concept of *Gaia,* James Lovelock argues that the relationship between life forms and the planet

Box 4.1 Beaver dams, sand dams and the Three Gorges Dam

Beavers are expert engineers, doing jobs also done by human beings when managing their habitat – coppicing, damming, putting in sluice gates, thinning woodland.'The damage they do pales into insignificance compared to the good they do for the ecosystem,' in the words of one expert.[1] The sand dam (invented by the Romans and currently used with some success in Africa) won a virtual prize in an Earthwatch debate in 2009.[2] Being small scale, the sand dam is perhaps analogous to the beaver dam, since, as well as providing a source of water for human beings during the dry season, it ends up creating a small-scale local ecology that has benefits beyond those enjoyed by the builder. A small (1–5m) concrete dam is built across a river and the up-river side of the dam, where a small reservoir of water forms, is filled with sand. The sand stays wet for a long time when water levels are low, acting as a filter and slowing evaporation, keeping a store of water in place by this means until the next rainfall.

When infrastructure is built in small scale it is more likely that a balance will be found between the builder and the environment. When in large scale, it is likely that environmental damage will result, but also that the structure will come under pressure from the environment. The construction of the Three Gorges Dam, China, began in 1992, and involved some 16 million tonnes of cement, creating a huge reservoir covering a tract of land almost as long as the UK. The huge change in water volumes in the river upstream and downstream of the dam has triggered landslides and the lower flow of the waters has made it difficult to flush pollution from the river. Plans to build an 'eco-buffer' along the banks of the river are under consideration.

is interdependent.[3] At its best, it is 'in balance' or (to use a rather dour economics term) in equilibrium, implying an element of self-regulation between the competing infrastructural needs of different organisms. The entire surface of the Earth including life can be thought of as a single, 'self-regulating entity' (Lovelock, 2006). Furthermore, it is self-regulating precisely because of the organisms upon its surface – hence, plant life helps stabilize the concentration of greenhouse gases, which in turn regulate temperature, which in turn keeps conditions such as temperature stable for plant life. Lovelock would find these pages far too human-centred, but his writings could not be more relevant than in the context of infrastructure, which can 'lock in' human impacts on the planet for many years.

What defines economic infrastructure?

So what is the human impact – or the economic infrastructure? Infrastructure investment is a very specific form of spending. Infrastructure tends to have a long life. Once infrastructure is built it can last a very long time indeed (one of the ancient world's Seven Wonders still exists today). Infrastructure also tends to be expensive, requiring significant capital investment up front. Very often, the economic return earned from infrastructure may start to accumulate only several years after the infrastructure has been acquired. Similarly, the environmental impact of infrastructure may only become apparent years after it has been in place.

Moreover, the performance of any given piece of infrastructure – whether financial or environmental – may (depending on precisely what it is needed for) depend on a steady investment in maintenance through its life. Infrastructure is therefore peculiarly dependent on borrowed money over the long run, but also dependent on the sound finances of the owner on a year-by-year basis.

The cost of sticking with what you know

Because infrastructure is long-lived, infrastructure inertia becomes important. Once a commitment is made to having a certain form of infrastructure in place, the cost of changing the system goes up. The history of railways in the UK provides an outstanding example of this process. In the mid-1960s, two Beeching Reports recommended a sweeping disinvestment in railway networks in the UK. As a consequence of recommendations in the report, one-quarter of the UK rail network was dismantled. Dr Beeching's so-called 'railway modernization scheme' was reportedly intended to cater for 'a growing population and freight movements increasingly served by alternative means of transport such as cars and aeroplanes'. By committing to the infrastructure necessary for cars and aeroplanes, at the expense of railways, the option of *returning* to a railway-based transport network today becomes prohibitively expensive, in economic terms. It also, of course, prejudices the consumer in favour of travel by car (given the limited alternatives, particularly for rural areas). Consumer demand for cars in the UK today is (in part) determined by the decisions of Dr Beeching 50 years ago.

Environmentally, the cost of *not* having a railway network today raises the price of the decisions of half a century ago. In France, rail travel has an extremely low-carbon footprint because it is fuelled by electricity from nuclear power. It may well be unfair to comment specifically on the decisions that followed from Beeching. A good network confers flexibility, and under-used branch lines (which appear to have been one of the targets of the report) would be unlikely to achieve this. However, with hindsight, maintaining more of a balance between rail and road transport (rather than shifting transport policy towards roads) would have been a higher-quality policy decision. Cutting railway infrastructure in the 1960s effectively cut off flexibility in shifting the balance between road and rail transport today.

Running up an environmental credit card bill

So far, we know infrastructure is expensive and that it generally requires economic credit to finance it. However, infrastructure has another bearing on the environment. The infrastructure that we have dictates what natural resources we consume – because it often can influence the local (economic) price of those resources.

Let us turn to Dr Beeching's realm of transport, once again. The consumer is generally a critical part of the transport infrastructure network – nothing symbolizes the modern consumer society better than a car. The relevant form of transport infrastructure, the road network, demonstrates the two-way dynamic between technological development and the shape of the infrastructure.

The (now conventional) road infrastructure is a by-product of the invention of the wheel, which created the need for smooth firm surfaces over longer distances. Gradually, the wheel was more effectively harnessed to more-advanced forms of energy – first to horses, and then to engines. The ability of wheeled transport to go longer and longer distances, expanding economic activity, and expanding populations, vastly expanded demand by removing former constraints, eventually leading to the proliferation of the road networks we know today.

It can be difficult to tease out the relative impacts of technology, economics and politics, but the recent history of road transport in China provides a convenient experimental 'control'. It was clearly not the invention of the wheel (thousands of years ago), nor the invention of the car (invented a century or so ago) that has driven the rapid expansion of Chinese motorway networks in recent decades, together with the rapid increase in car ownership. These pre-existing conditions were necessary but not sufficient for an expansion of road networks. Today, Chinese cities are gridlocked and the air is blue with exhaust fumes. The cause of this is a road infrastructure that mis-prices road use, for a population that can afford to use the (too cheap) roads.

The road network is a costly resource. In modern times it is also a scarce resource. The UK Department for Transport's 'Eddington Transport Study' (published in December 2006) reported that, in the UK alone, the transport infrastructure supports 61 billion journeys a year. The report calculated that a 5 per cent reduction in travel time for business travel on roads could generate £2.5 billion in savings for the economy (time is money, after all). At the same time, the research predicted 13 per cent of traffic would suffer stop-start conditions by 2025.[4]

So far, policy suggestions designed to deal with congestion have tended to be superficial rather than structural: controlled use of the motorway hard shoulder at peak travel times, investment in local transport and limited congestion charging. None of these (bar perhaps the last) addresses the possibility that the transport system is the way it is because of the structure of the underlying infrastructure, which in turn grew up in the way it did because transport costs were subsidized. 'As road space is a valuable and scarce resource it is natural that economists should argue that it should be rationed by price … If road users paid the true social cost of transport, perhaps urban geography, commuting patterns and even the sizes of towns would be radically different from the present' (David Newbery, cited in Jenkinson, 1996, p141).

What is happening here is that the existence of infrastructure (roads) and the pricing of that infrastructure (free), perhaps combined with the limited number of alternatives (lack of railway investment), is combining to prejudice consumer decisions. Consumers buy cars because all these factors come together at once.

The descriptions of traffic conditions in Asian cities suggest that exactly the same model is being followed. New road networks are needed because, if they are not provided, population growth and economic growth will simply lead to more and more congestion. Infrastructure (a plentiful supply of roads) generates a low price (absent road tolls), which gives consumers an incentive to use cars that consume scarce environmental resources (oil). One could suggest that the road infrastructure does not need to be upgraded per se (although many motoring organizations would dispute that). However, the existence of the road infrastructure incurs an environmental credit cost, in the absence of pricing or other controls.

Transport is just a small instance of the way in which the built infrastructure shapes the way in which people draw down environmental credit. People have always gathered together to make things, or to trade, but perhaps never more so than in the 20th and 21st centuries, and an entire infrastructure has grown up around modern working practices. Productive activity is housed in the modern built environment and relies on a transport infrastructure to take people from home to work and back. The way people live when they are at home has always been shaped by their work. In modern times, many elements of housework have been delegated – cooking to the large food manufacturer, and washing and cleaning to machinery driven by energy. Infrastructure clusters in several areas – the built environment and its service components; transport and storage; the manufacturing environment and energy; water and wastewater – support this way of life and have also been shaped by this way of life. Once again, the question is whether these working and living patterns would have developed had infrastructure usage included embedded environmental and social costs.

It is important to recognize that the shape of the built infrastructure regulates natural resource usage as well as the volume of service traffic for which it is intended. Hence, the extent of the road network sets the boundary to the number of car journeys and energy-generating capacity sets a limit to how much power can be used, and in both cases available capacity determines the extent of the environmental impact. Depending on how it is designed it can potentially set limits to usage that are too high for sustainability. As this suggests, one solution to the problems of environmental impact is to change the infrastructure, and that is an investment issue. An economist would argue that this is why we need economists, and an environmentalist would point out that economists did not do a great job of things first time around.

More broadly, the existing infrastructure has been shaped by and also shapes living patterns: work, travel, leisure, communication. The investment challenge is twofold: to reshape infrastructure so that it results in living and working patterns within the limits of planetary resources; and to change the way existing infrastructure is used in such a way that its direct human environmental footprint is minimized.

The tension of economic infrastructure in a changing environment

Here, then, is where economics and environment go head to head. Putting infrastructure in place requires long-term capital commitments, generally funded by financial credit. Due to this, there is a strong economic desire to keep the existing body of infrastructure *in situ*. There is little (economic) advantage from abandoning a power station after 15 years if the expected life of the power station is 50 years, for instance.

The challenge is that the perceived external environment at the time an infrastructure project is built may be very different from the external environment that exists 15 or more years later: Dr Beeching was not aware of carbon footprints, greenhouse gases or peak oil, and so came to a perfectly rational decision (from his perspective). Confronted by today's environmental credit crunch, with its implications for energy demand, we may have cause to regret the decisions of the 1960s, considering them ill thought-out.

The dark cloud of the financial credit crunch may have a silver lining for infrastructure investment, if one consequence of constraint is that better decisions are reached about what infrastructure to build and where to put it. Against that optimism there is a danger that the financial credit crunch encourages infrastructure investment that is shaped by what is already there, perpetuating resource intensive living patterns. After all, under (financial) duress, it is all too easy to be blinkered by present-day constraints when planning the future. This is something we highlight later in this chapter, when we look at the role that government has to play in the wake of the financial credit crunch.

Infrastructure and the twin credit crunches

The credit crunch acts in three ways to impact the level of investment. For private investment this is through the price, and through the amount of investment funds that are available. The two ideas are related, but (as we shall see) are still worth treating separately. Added to the investment in private infrastructure, there is the issue of public investment – and the credit crunch has certainly altered the scope of public investment in the medium term.

Affording infrastructure in the wake of the financial credit crunch

The economics

The financial credit crunch has made it more expensive to borrow, thus more expensive to invest. That renders investment in new infrastructure less likely. What is the importance of the price of investment, and how has it gone up?

Interest rates and borrowing amounts

Consider someone who took out a floating rate mortgage circa 1990. Around that time, at least in the UK, mortgage interest rates were commonly around 15 per cent. This meant that the interest payments (or debt servicing) would be a large chunk of a mortgage payer's income. The mortgage payer would only be able to afford to borrow if they could afford to meet the interest payments – and as interest payments were so high, this was a natural constraint on how much someone could borrow. A mortgage of £75,000 would perhaps be within the means of someone earning £25,000 (before tax) per year. The annual interest payments, at nearly £11,000, would eat up a very large proportion of the mortgage holder's income, of course. A mortgage of £100,000 would be quite beyond their means.

If mortgage rates were lower, the position changes. Say that mortgage rates were 7 per cent, and borrowers believed that they were going to stay around that level for the life of the mortgage. The £75,000 mortgage can be funded with consummate ease. At 7 per cent rates, the interest charge is barely over £5000. On an income of £25,000 this is easily managed. Indeed, it is so easily managed, that there is no reason for the borrower to stop at a £75,000 mortgage. Why not go a little further? After all, the interest charges on a £100,000 mortgage will only be £7000 – lower than they were with a £75,000 mortgage in the bad old days of 15 per cent interest rates. It would be foolish to stop at £100,000. As long as the borrower is sure that the interest rate has gone down, the amount borrowed can go up (because it is more affordable).

Of course, the amount borrowed could stay the same and simply be repaid in a shorter period of time. Strangely, borrowers seem not to opt for this alternative terribly often.

If the cost of borrowing goes down, people (companies, society at large) can afford to borrow more money on the same income. If the price of borrowing goes up, then less money can be borrowed and less investment will take place.

Lowering the price of borrowing

The past 20 years have seen a steady reduction in the cost of borrowing money. From around 1990 onwards, the price of borrowing for individuals, for companies and for governments generally came down. Why did this happen? It happened because risks were reduced.

Lenders want a return on their investment (a profit) of course. That is the whole reason for lending. But in addition to their profit, the lender will want to be compensated for uncertainties or risks. What happened during the 1990s and the early years of this century was a decline in risks. More accurately, some risks declined and some risks did not decline but banks and other lenders thought that they had declined. Whether the decline in risk was real or imagined, the result was a lower cost of borrowing. If the risk was lower, the compensation required to persuade a lender to take the risk was lower, so – basically – interest rates fell.

The result was cheaper borrowing for all – governments, corporations and individuals were able to indulge in a borrowing binge. Borrowing was cheaper, and one could borrow a higher proportion of one's income (increased leverage).

Raising the price of borrowing again

So, where is the problem with the price of credit today? The problem is not the official, central bank cost of borrowing money. The problem today is that banks and other lenders have increased the cost of borrowing money in the real world, as perceived risks have risen. Indeed, in some cases, the cost of borrowing money has increased infinitely (because no money is available 'at any price' – something we shall cover later).

Risks that never really went away have revealed themselves and risks that had been minimized have suddenly shown new vigour. Lenders have suddenly come to realize that lending someone five times their income, when the only proof of their income is what the borrower claims it to be, is not necessarily the low-risk strategy that the lenders believed it to be. In fact, some lenders have come to the conclusion that it is quite a high-risk strategy.

Other risks have spun out of the financial credit crunch. If global growth is going to suffer, a new risk premium is required. The rise in trade protectionism raises concerns about the future direction of globalization – another risk that requires compensation. There is another, new risk that is very specific to the aftermath of the financial credit crisis. For many (especially individuals and small businesses), banks remain a key part of borrowing.[5] In the aftermath of the credit crisis, banks are being subject to more and more regulation. That raises the cost of borrowing.

Banks, borrowers, regulators and rates

There is little doubt that more regulation is required. The fact that banks collectively took bad decisions was not a conspiracy – it was at least in part a consequence of the failure of regulation. However, regulation, even required regulation, entails costs. And someone must pay for those costs.

Think about a bank battered by adverse market forces, forced to rethink its lending philosophy and adjust its perceptions of risk. After all of that, governments then descend with a sheaf of new regulations. Banks need to understand and implement all of these regulations. They must make sure that they comply with the regulatory regime – not necessarily because the regulations are sensible (politicians tend to make sensible regulations only by accident), but because the financial and reputational damage of non-compliance could be severe. Faced with such a barrage, there can be only one solution: lawyers.

Lawyers cost money. Generally, lawyers cost a lot of money. The average annual compensation for a lawyer in the US in 2009 was $129,020. The average compensation for an economist, in contrast, was $96,320, and the environmental scientist could only command an even more modest $67,360. This is what economists call 'market failure' – an instance where the free market fails to reflect the true relative worth of different groups in society.[6] Banks will pass on the cost to their customers – who are borrowers. The borrowers will end up paying a price for greater bank regulation through increased interest costs.

Pricing and infrastructure

What the credit crunch has done, therefore, is raise the cost of borrowing money. With a higher cost of borrowing money, investment is likely to be lower in the future. What the credit crunch boils down to is the harsh fact that the price of investing in infrastructure is going up, so the amount of investment in infrastructure is likely to be lower.

The environment

A higher cost of borrowing entails less infrastructure construction in the future. Is this environmentally a bad thing? Net, it is. However, it is not a universal negative.

Infrastructure apparently without purpose – human folly

Not all built infrastructure has a clear purpose. Many large country houses built in the British Isles in the 1700s and 1800s boast a 'folly'; an ornamental building with no practical purpose (other than visibly to proclaim the wealth of the owner of the land).

Follies did not die out with the decline of British country house building. Modern society is littered with infrastructure that appears to be built primarily to show off economic prowess or extreme wealth, with no thought of economics or the environment. What else can possibly explain the appearance of the famous Palm Islands in Dubai? What history will make of them remains to be seen; whether they will last enough for history to judge them (given the need for constant dredging to keep the islands in shape) is a moot point. Consider the race for the tallest building. The Empire State Building held the lead for 42 years. The Petronas Twin Towers held the position for six, as did their successor, Taipei 101, before Dubai (again) reached the pinnacle with the Burj Khalifa in 2010 (a tower whose name owes much to the consequences of the financial credit crunch. It was named after the ruler of Abu Dhabi, who had come to the financial aid of Dubai following the collapse of the Dubai property market).

Deterring economic and environmental follies is a benefit of the financial credit crunch. The froth of excess has been blown away by the chill winds of higher risk and the associated higher cost of capital. But, of course, the benefits of fewer follies are outweighed by more serious costs. Investment in environmentally necessary infrastructure is subject to the same restrictions as follies.

Prolonging the environmental credit crunch

Some pieces of the built infrastructure will have an impact upon the environmental profile of others. The energy sector is a case in point. We examined energy in more detail in the previous chapter, but a simple case study demonstrates the environmental importance of rising infrastructure costs.

As things stand, at the start of the 21st century, the energy infrastructure (both liquid fuels and electricity) is heavily dependent on fossil fuels. The question is whether 'green' energy would be more developed than it is had the price of fossil fuels fully incorporated their environmental costs. It seems reasonable to suppose solar energy would be far more widely used than it is currently. However, now that environmental costs are at least better understood (if not necessarily fully reflected in market prices), why is there not a switch to solar power?

The key solar power generation technology is actually concentrated solar power (CSP). The technology is known to every school child: focus the sun's rays on one spot through a magnifying glass and this produces enough heat to start a fire. Using parabolic mirrors to the same end, enough heat can be generated to drive steam turbines.

The reason why this technology has not developed lies in an infrastructure problem. To move the electricity generated from such technology (positioned in hot deserts), high-tech transmission networks are required – so-called High Voltage Direct Current (HVDC) networks.[7] HVDC networks do not yet exist everywhere they are needed – they require construction, which means they require funding. However, as the cost of funding goes up, the economic cost of new generation and transmission capacity directly related to large-scale solar power goes up. Policymakers end up comparing the cost of fossil fuel power generation (with existing infrastructure) to the cost of solar power generation (requiring new infrastructure at a higher funding cost). The comparison becomes steadily less favourable (in economic terms) in a climate of rising interest rates.

China, which has abundant hot desert land containing potentially significant sources of solar and wind power, also has the (domestic) problem of very long distances and very large cities requiring powerful point sources of electricity. However, it is perhaps the space to watch. China lacks the infrastructure needed to satisfy its future energy requirements – *any* additional power demand will require additional investment. This then is not a comparison of existing infrastructure with new (costly) infrastructure, but a comparison of different forms of infrastructure, either form of which will incur higher capital costs. Here is a possible area of the environmental credit crunch being managed appropriately – but it is an instance that is specific to the emerging markets and not to the established economies of the Organisation for Economic Co-operation and Development (OECD).

To return once more to the transport infrastructure – insufficient funding (a direct consequence of the unaffordable cost of capital) has meant insufficient investment in rail infrastructure that generates an extreme risk in some economies: in 2008, on average, 17 people died every weekday on the suburban rail network of Mumbai, India. A visit to the central railway station in Mumbai readily explains why. Trains arrive in the station not just packed to the ceiling but beyond. People holding onto the doors hang outside the train, and sit on top of the train, a few inches from high-voltage wires. The roads are so congested that apparently people will risk death to get onto another form of transport, in order to get to work. In 2008, 853 people fell (or were pushed) to their deaths from overcrowded carriages. A further 41 were bludgeoned to death by trackside poles as they hung from the train, and 21 were electrocuted.[8] The cost of capital in India (since independence) has been so high as to deter additional investment in the transport network, and thus freezes large amounts of the rail infrastructure in their state circa 1945.

The amount of money available

The economics

Infrastructure tends to require long-term financing to make it viable. Infrastructure projects are rarely short term. This means that the pool of money available for infrastructure spending is constrained by the amount of money that lenders are prepared to put to work over the longer term.

Free market economics tells us that the supply of long-term money (lenders' funds) will be matched to the demand for money (infrastructure investment) at the free market price (the interest rate). The price of long-term money is increasing as a result of risks and other costs – is it worth exploring the pool of money any further?

There are a couple of reasons why thinking about the amount of money available may be helpful. Of course, everything could be viewed through the prism of the price mechanism, but this is not necessarily the most helpful way of understanding how the credit crunch has had such a pervasive impact on infrastructure.

Not at any price

The first issue with the supply of credit is that, at least for some, the economic events of the past couple of years have now cut off the supply of credit to some sections of society. As the perception of risk waned from the 1990s, so lenders were willing to lend to ventures that were previously considered 'beyond the pale'. Of course, not all of this lending was wise (with the benefit of hindsight). However, in reaction to the events of the past couple of years, lenders have retrenched. High-risk borrowers will now not be able to borrow (and threaten the newly cleaned balance sheets of the lending community). This is not a simple case of not being able to borrow at previously attractive rates of interest. This is a case of not being able to borrow at any economically viable rate of interest.

Evidence of this can be seen, in extremis, in the consumer sector. Pawnbrokers typically charge interest rates of around 100 per cent per year or more. Denied access to *any* credit from mainstream sources, families have increasingly turned to pawnbrokers as a source of capital. Pawnbrokers' profits have soared. What this is indicating is that mainstream lenders have become so risk-averse as to deny credit (and thus the ability to invest in infrastructure) to a section of society – be that household or corporate. The family wanting to replace oil-fired heating with a wood-burning stove will not get the credit they require. The company seeking to update its machinery for a more energy-efficient model is told that it represents too great a risk.

We must have discipline

The second area where the amount of credit is being denied is via the reregulation of the financial system. Again, we are treading on contentious ground. There is no denying that reregulation, and better regulation, is required. However, governments are not necessarily regulating effectively, and this ineffectiveness may unnecessarily shrink the amount of money that can be spent on infrastructure.

One issue with regulation is the amount of capital banks have on their balance sheets. The arguments surrounding this can generally only be understood by accountants (which describes neither of the authors of this book). The bottom line is that banks across the world are being asked to have more cash tucked away in reserve, just in case things go wrong again in the future. This new discipline provides security – and can be seen as a perfectly sensible strategy. However, the increase in capital requirements for banks also means that less money is available for lending (if it is being held as cash, it cannot be used to pay for a new power plant or a new wood-burning stove).

If politicians get the capital requirements wrong – and insist that banks hold too much money in reserve – there will be less cash available for infrastructure spending.

The issue with the amount of capital available is not, therefore, so much that the credit crisis has led to a drying up of cash. More, it is that any misguided regulation that may be imposed will reduce the amount of cash available by more than it should. There is no denying that some regulation is required. There is good reason to question the extremes to which regulation may go.

The environment

A rather extreme instance of the importance of available capital to the environment can be seen through the management of water resources. Five millennia ago, the Egyptians began to master the art of hydraulics to control the water from large rivers for irrigation in large scale (and in so doing triggered an agricultural revolution). In contrast, India and Bangladesh are still dependent on the natural regularity of the monsoon season and have not attempted to soften the impact of the extreme swings in weather conditions with large-scale built infrastructure such as dam systems. Despite the five millennia of knowledge, there is simply not the capital to implement the projects that would store water for use after the season, nor to provide defences to prevent the flooding, displacement and deaths that often seem to accompany the monsoon season.

This is a 'chicken or egg' situation. It is difficult to define whether the lack of infrastructure investment is driven by low levels of economic wealth, or whether growth is held in check by this lack of infrastructure investment. However, it is clear that the absence of capital (at any price) in India and Bangladesh is a limiting factor on managing environmental risk.

The role of politics in regulating and restricting investment is environmentally significant. However, government can also have a very direct impact on the environment through the way that it influences infrastructure.

The government and infrastructure

Government is back, no question. The scrapes and bruises of the credit crisis caused the financial sector to run shrieking into the embrace of the 'nanny state', crying for sticking plasters, antiseptic creams and a lollipop for being brave. The weakening of economic growth led to fiscal stimulus programmes across the OECD economies and emerging markets (China allowed itself no stint in this regard). If the consumer was suddenly shy, and companies were concerned about the outlook, it was down to the government to spend.

In addition, governments have seen their tax revenues dwindle. The assumptions of high rates of economic growth have had to be cut – and cutting growth estimates means cutting tax revenue growth forecasts.

The financial credit crisis has therefore had two consequences for government and infrastructure. The first is that the government is playing a larger role in the economy. The second is that the government has less money.

The economics (part 1): A bigger Big Brother

What does a larger government involvement in the economy mean for infrastructure? In one sense, it is potentially an environmentally positive development. Governments are less shackled by the market forces of profit and loss. Governments can indulge in 'cost-benefit analysis', which includes social and environmental consequences in their calculations. The result is that an increase in the role of government may lead to more environmentally positive infrastructure.

Consider one of the more high-profile instances of government support for the US economy – the 'cash for clunkers' programme. (This was formally known as the Car Allowance Rebate System and officially began on 1 July 2009). This government spending encouraged the US consumer to buy modern, more fuel-efficient cars, trading in their old and less fuel-efficient cars. The government is playing a larger role in the economy (by influencing the effective price of transport infrastructure for households), and is seeking to achieve an environmental outcome as a result.

The environment (part 1): The success and failure of government

On the surface, the US cash for clunkers scheme is a success. Some 700,000 cars were purchased (stimulating economic activity). The cars that were purchased under the scheme had a fuel consumption that was an average 9.1 miles per (US) gallon better than the cars that were scrapped under the scheme. Moreover, 80 per cent of the cars scrapped were pick-ups or SUVs (the more fuel-inefficient form of household autos), while nearly 60 per cent of purchases were passenger cars. Consumers were 'downsizing'. The back of an envelope calculation is that a little more than 3.8 million barrels of oil were 'saved' as a result of the upgrade.[9]

However, the bi-focal outlook of the government is a complication. The US government (and other governments around the world with similar schemes) undoubtedly received a warm glow of environmental satisfaction as they expanded their role in the economy. However, the other objective was, patently, to stimulate domestic demand and support the local economy. The US cash for clunkers programme was partly (probably mainly) motivated by the dire state of the US car manufacturing industry and the need to find someone willing to buy American-made cars. Given that the rest of the world has scant interest in US-produced autos, the US consumer needed to be stimulated to purchase.

Why is this bad? Of itself, it is not 'bad' – but it may not be environmentally optimal. Governments, when they provide stimulus to an economy, have to work with the existing economic structures. Rather than persuade people to find alternative environmentally friendly means of transport, the US government has to encourage them to buy new cars if it wishes to save the life of the US car industry and the jobs associated with it. Similarly, Europe's mooted ban on imported solar panels from China says more about domestic economic priorities (preserving domestic jobs) than either domestic or global environmental priorities (maximizing the use of solar panels).

As we do not live in a perfect world (a world where economists would make all the decisions except for the ones necessarily made by environmentalists), political reality means that government spending and stimulus is likely to focus on job creation or job preservation. This means that *existing* industries and *existing* technologies are likely to be favoured – there are sunk costs, which cannot be cast aside when employment is such a priority.

Politically, if the short-term economic outlook clashes with the long-term environmental outlook, bigger government is far more likely to favour the former rather than the latter. That is the nature of policy stimulus. There are votes in saving Detroit. There are likely to be far fewer votes in encouraging the import of foreign solar panels.

The economics (part 2): Can you spare Big Brother a dime?

The explosion of government budget deficits has led to a great deal of sensational comment. Debt levels have risen sharply in most OECD economies and a number of emerging markets as well. Although not yet comparable to the position of 1945, the stock of debt outstanding is certainly significant.

This increase in debt very often represents a socializing of private sector debt. Governments have effectively helped the private sector reduce its debt burden by offering fiscal stimulus. This is an intergenerational transfer of debt. Americans, needing to reduce their credit card bill, are offered a tax cut by the government. That tax cut is then used to pay down the debt outstanding (in this sense, the tax cut acts not as a stimulus, but as an anti-depressant for the economy). How is the tax cut financed? By borrowing, which in turn will be paid for either through higher tax revenues or lower government spending at a later date. Thus, the next generation receives the bill. One's children are effectively paying for one's credit card.

If governments wish to reduce the debt burden that now exists, or if markets force governments to take steps to reduce their debt burdens, then the inevitable bias is for a lower level of government spending than would otherwise be the case. This argues that governments will play a reduced role in providing infrastructure in the coming decade. It also suggests that maintenance of existing infrastructure may be lessened (that, for instance, was one of the consequences of the need to reduce the UK's national debt in the latter half of the 20th century).

In fact, the argument goes beyond reduced government spending. If a government is acting under fiscal constraint, it is likely to be constrained in how it uses its tax policy as an incentive mechanism. Governments may be unable to offer tax incentives to encourage investment in environmentally positive infrastructure. Where there are environmental benefits from investments (positive externalities, as economists like to call them), which are not reflected by market forces, tax incentives can be beneficial.[10] If governments cannot offer tax incentives (or, indeed, subsidies) that are equal to the value of the positive externalities, then there will be less investment in that form of infrastructure than is economically or environmentally optimal.

The environment (part 2): Infrastructure the government cannot afford

Government is, if not indispensible to environmentally friendly infrastructure, at least highly helpful to the cause. Environmental externalities are rarely captured well by market forces. Economic textbooks often illustrate external costs that are not captured by market forces through the example of a factory issuing smoke from its chimneys, positioned next to a laundry – environmental externalities and the failure of the market to deal with them is pretty much a hackneyed concept in the world of economics. Governments can assist in the process of socializing the cost of the environmental investment and making it more attractive. What the credit crunch has done is distort and hinder that process.

Once infrastructure is in place, it must be maintained. Steven Solomon (2010) observes that the 'economic and cultural dynamism' of any society is signalled by the state of its waterworks, defined as any aspect of water-related infrastructure, from water provision to flood defence. It is likely that the same could be said of almost any infrastructure that supports the needs of the relevant civilization. Hence, the earthquake readiness (or not) of the built infrastructure, the state of roads, the reliability and safety of railways, the state of public buildings and housing, and the preparedness of the energy infrastructure, may all at some stage act as a bellwether in respect of the broader health of a society and its economy. The signals can sometimes be contradictory – what is to be made of the careful plans for the Thames Barrier alongside a Victorian water infrastructure that leaks 60 per cent of its contents, a drainage system that puts raw sewage into London's greatest river when levels of rainfall overwhelm the drains, and government plans to build all over a critical piece of natural infrastructure – the floodplains of the River Thames?

The point is that cash-constrained governments in the era of the financial credit crunch may not be able to build or maintain infrastructure that the environmental credit crunch requires. Indeed, the failure to maintain the existing stock of infrastructure can actually compound the environmental credit crunch – leaking water systems do not facilitate the environmentally prudent management of water resources, fairly self-evidently.

Box 4.2 Canals and manias

As ever, the ghosts of history continue to haunt today's economist. Before the internet, before motorways, before railways, there were canals. Canal building captures many of the problems that plague infrastructure, finance and the environment.

Canals are nothing new. The Grand Canal of China, completed in the early 7th century AD, has still not been surpassed as the longest canal in the world. It starts at Beijing and extends south to Hangzhou (a staggering 1103 miles in length). This canal connects China's two major water systems, creating the biggest inland waterway network in the world. It facilitated economic development: 'By bridging China's North/South hydrological divide, [the Grand Canal] synergized the natural and human resources of the two diverse geographical zones to help launch China's brilliant medieval golden age' (Solomon, 2010). It was also, by and large, dependent on economic abundance – not to finance it (this was ancient China), but to build it. Cheap, available labour made the canal possible. The absence of economic resources in the 19th century contributed to the deterioration of the canal, contributing to the floods that ravaged the lower Yangtze in 1849.

The UK in the 18th and early 19th centuries offers a more direct analogy to the interaction of infrastructure and financial credit. The second wave of canal building came in the late 18th century, as a means of transporting goods over large distances far more cheaply than had ever occurred before. The motive was consumer demand – the Duke of Bridgewater kicked the whole thing off, as he sought to meet consumers' demand for coal by linking his mines to Manchester with a canal.

Canals may have been derived from consumer demands, as being economically more efficient than land transport (until the advent of railways). However, canals were environmentally more efficient too: the Montgomeryshire Canal Company altruistically declared that it was not in it for the profit, but to accomplish (among other things) 'the Extension of Agriculture, the Reduction of Horses'. Horses are energy consuming, just like cars – greater equine efficiency means greater environmental efficiency.

In the late 18th century, a 'canal mania' took hold and companies found it easy to raise capital with which to build the required infrastructure. Prior to the mania, a canal company that borrowed money in the 17th century would generally pay 6 per cent for the privilege. In the early part of the 18th century interest rates were 5 per cent, while by the middle of the century most companies paid 4.5 per cent or less. In fact, the main mechanism for raising money was share issues and, here too, the availability of credit was plentiful. A canal share worth £10 in 1775 would trade for £200 18 years later. (Interestingly, banks did not participate in canal financing. The only notable bank finance was from Bath City Bank, which invested in the shares of the Salisbury and Southampton Canal. Both the bank and canal failed in 1793.) Canal building boomed, and then in 1793 the bubble burst. Companies could not raise capital on such easy terms – nearly all borrowing in the latter 1790s was at 5 per cent. Share prices collapsed. Critically, this financial credit crunch restricted investment in infrastructure. The Herefordshire and Gloucestershire Canal, for instance, was abandoned (half finished) for want of money in 1798. Investment in that particular infrastructure did not restart for nearly 40 years.

There was another bout of canal mania in the UK in the 1820s, but smaller-scale canals were being superseded by railways by the 1830s (an instance, like Dr Beeching 130 years later, of a failure to anticipate shifts in infrastructure). Throughout the mania, the environmental and economic efficiency was contingent on the provision of financial credit, every step of the way.[11]

The financial credit crunch, the environmental credit crunch and infrastructure

A sound infrastructure delivers enough, but not too much, of the required service, with minimal impact upon the ecosystem that supports it. Infrastructure should, in an ideal world, be designed in such a way that it is at one and the same time protective of the environment and yet also delivers the required service reliably, safely and efficiently. Among other things, this means that the amount and positioning of any given infrastructure needs to be optimal in respect of human need. Too much infrastructure in place, delivering resources such as water, energy and transport too freely, is likely to lead to their over-use.

Eventually, this problem will become an environmental credit problem, because the ready availability of any given resource will tend to shape other aspects of the built infrastructure, which will be designed with the assumption of cheap availability of the relevant raw resource in mind. This then leads to a further misallocation of environmental resource, causing a further environmental credit crunch – and so on in a vicious circle. Generalizing, the current state of built infrastructures is that they encourage the over-use of limited resources because of the way they have developed over the years through a mix of accident and design.

The very existence of Earth Overshoot Day, which was highlighted in the Preface to this book, tells us that the existing man-made infrastructure of planet Earth is not doing its job properly. We need to change and upgrade the infrastructure to reduce (hopefully offset) the environmental credit crunch. The problem is that the financial credit crunch does not facilitate that process. If anything, Earth Overshoot Day is being brought forwards, rather than postponed.

The environmental credit crunch demands more investment in the coming years. The financial credit crunch, through a higher cost of capital, denies us the money with which to undertake that investment. The political preoccupation with the financial credit crunch (and its electoral consequences) means that any solution – even those tastefully veneered with the respectability of environmentalism – will focus on supporting the industrial and infrastructure status quo. Even if efficiencies increase, the step-changes required to deal with the consequences of the environmental credit crunch are unlikely to be undertaken (for the foreseeable future) because of the financial credit crunch.

Existing infrastructure is an impediment to change in the way resources such as water, energy and land are used. Infrastructure investment could be an opportunity to change the way scarce resources are used. Not investing in infrastructure (allowing existing infrastructures to become increasingly congested) may be one route to constraining use, but as this chapter suggests, this could be bad for the economy, or bad for the environment, or both.

In economic infrastructure and environmental infrastructure terms, this was the worst possible time to have a financial credit crunch.

Notes

1 Jenny Holden, Beaver Trial Field Officer, cited in *The Independent*, 15 July 2009.
2 Earthwatch is an international environmental charity founded in 1971. See www.earthwatch.org.
3 See, for example, Lovelock (2006).
4 Details of the Eddington Transport Study are available at http://webarchive.nationalarchives.gov.uk/+/http://www.dft.gov.uk/about/strategy/transportstrategy/eddingtonstudy/.
5 Consumer spending makes up roughly 65–70 per cent of economic activity. To satisfy that consumer demand, companies provide goods and services, of course. Around 60 per cent of corporate activity is provided by small businesses – what Americans call 'mom and pop stores', and what the Germans call (with a somewhat more dignified expression) the 'Mittelstand'.
6 This horrifying data comes courtesy of the US Bureau of Labor Statistics, the May 2009 National Occupational Employment and Wage Estimates data. Lawyers have occupation group number 23-1011, economists are 19-3011 and environmental scientists 19-2041.
7 The usual objections to this idea are that, even if it were technically possible, it is politically impossible since it would require the building of high voltage networks across politically sensitive regions. For some reason, something that was not an issue in the context of oil and gas is a show-stopper when it comes to CSP.
8 This story was reported in *The Times* (UK) newspaper on 1 April 2009 as 'Mumbai's overcrowded trains kill 17 people every day'.
9 The success of the policy was proclaimed by US Transport Secretary Ray LaHood, cited on www.climateprogress.org.
10 One example in the UK is the tax incentives that surround forestry projects – which generate environmental, recreational and aesthetic benefits that are rarely captured in the market price for timber. However, there is some question as to whether the tax advantages of investing in forestry exceed the current calculation of the positive externalities.
11 Full details on this episode, the canal mania, are to be found in Ward (1974).

References

Jenkinson, T. (ed) (1996) *Readings in Microeconomics*, Oxford University Press, Oxford
Lovelock, J (2006) *The Revenge of Gaia*, Allen Lane, London
Solomon, S. (2010) *Water*, HarperCollins, New York
Ward, J. (1974) *The Finance of Canal Building in Eighteenth-Century England*, Oxford University Press, Oxford

CHAPTER 5

Housing: The Canary in the Coal Mine

I been laid off from work,
My rent is due,
My kids all need
Brand new shoes,
So I went to the bank
To see what they could do.
They said 'son – looks like bad luck
Got a hold on you'
Money's too tight to mention
(J. Valentine and W. Valentine, Windswept Pacific Music Ltd)

Back in 2007, there was no such thing as a financial credit crunch. The problems of the global economy were confined to the housing market, and to the US housing market at that. Moreover, the middle classes could feel complacent – even in the US. This was not a general mortgage crisis. This was a sub-prime crisis (and how could anyone middle class ever be considered to be sub-prime?).

Sadly, the castle of complacency inhabited by the middle class soon surrendered to the besieging forces of global credit. From problems with sub-prime mortgages, the world economy has cycled through greater concerns regarding consumer credit (in all its various forms), to problems in the banking system and (ultimately) to issues with sovereign credit. The changed credit environment will impact the supply of housing, from both the private and the public sector. It will also change the way in which we demand housing. Housing demand is relatively simply resolved in the Organisation for Economic Co-operation and Development (OECD) economies. One lives in a home one owns. Alternatively, one rents a home. If neither of those are options, there is the possibility of social housing in some economies. The financial credit crunch potentially impacts each of these three forms of housing demand. As the financial credit crunch has a bearing on the way we live now (through influencing where we live now), it must surely have environmental consequences.

The housing sector is a profligate user of resources. With a few noteworthy exceptions, the way homes are built now gives present and future residents little choice in the matter of how much water, energy and other natural resources they use. Grey water spirals inevitably out of bath or shower into the wastewater system, rainwater pours unnecessarily down the drain, inefficient equipment recklessly consumes energy, and householders generously (and uselessly) send warmed (or cooled) air outside. Even the cry of 'location, location, location' (beloved of estate agents) has environmental consequences: location means that instead of travelling under our own steam we are, far too often, forced to rely on fossil-fuel based energy to take us where we need to go. The question is whether the financial credit crunch will effect enough of a change to how we live and where we live in such a way as to avoid the approaching environmental crunch. Putting this in another way, can we release the housing sector from the grip of the environmental mortgage it is currently locked into and will, one day, have to make good?

Location, location, location

Location is everything when it comes to property. How constrained land is, as a resource, depends on where it is, and this in turn has a strong influence on its worth. As an extreme example, in the Japanese real estate bubble in the 1980s, the small area of land under the Imperial Palace in Tokyo was estimated to be worth the entire state of California. Seemingly, the more crowded land is, the more valuable it is, and when land is very crowded, two things become extremely important from the perspectives of supply and demand alike: technology, and urban planning and regulation.

Technological developments in construction can make a big difference as to what can be built where. Chicago, the 'Windy City', used to be famous for its skyscrapers, but these days every new Asian city worth its salt has a building competing to be the highest in the world. The problem of urban sprawl, which happens as cities move to accommodate an increasing number of new arrivals, can be resolved to an extent by making sure all brownfield land is properly used and, by building vertically, thereby increasing population density. In a small, crowded country such as the UK, agriculture, the ecosystem, leisure amenities, the transport system and housing all compete for space and so planning regulations determine who can build what where. Data published by the UK Valuation Office cite survey-based land valuations in the order of £20,000 per hectare for agricultural land based in the UK's south-eastern regions, versus a range of £1.4–4 million per hectare for residential land. The huge premium for residential land is of course ultimately determined by the UK system of planning regulations as well as the nearness of the relevant infrastructure and general amenity value.

Amenity value, for any given property, will be a complex bundle of inputs, and, depending on where it is situated, building design and engineering ('technology') may also have a bearing on the valuation. As an example of buildings technology the Earthship (a self sustaining home) suggests buildings technology is likely to be more important in an urban environment.[1] As soon as people are living on top of each other in large numbers, they (inevitably) create environmental problems for each other. The denser the population, the greater the need for engineered solutions to ventilation; heating and cooling; water and energy provision; waste disposal; and transit.

Technology, however, cannot solve all problems, particularly when they are not anticipated and built into building design. One of the problems with large concentrations of concrete, glass and asphalt is the so-called urban heat island effect. Large cities tend to be slightly warmer than average because of the tendency of the materials used to build cities to act as a storage heater, as well as the cumulative effects of the heat thrown out by cooling systems and other equipment.

There is no shortage of examples of high-temperature living to suggest that the heat can be dealt with through design. Farmhouses with thick clay walls and small windows in the Gobi afford adequate shelter from the heat and the families within manage to continue to scrape a living by adapting to the physical conditions. The key point is that construction techniques were shaped by local conditions – which include temperature as well as available materials. The problem for cities suffering from heat island effects in the context of higher average temperatures is that they may not always be designed to keep their residents comfortable because they were designed for a different set of conditions. If the problem becomes extreme, this environmental issue will eventually affect demand unless buildings can be adapted in a cost-effective manner to deal with it.

Some cities are quite low-lying and this raises the spectre of flood risk. The UK Environment Agency provides a raft of advice to the householder about what to do to protect his or her home from flood. The ultimate protection is, of course, not to live where flooding is likely, and this makes major housing developments such as the Thames Gateway a puzzling idea. As far back as 2005, a report of a London government body (the GLA) suggested that 1.25 million people could already be at risk from flooding by the Thames and furthermore suggested that a major flood in the Thames Gateway could cost as much as £12 billion.[2]

Of course, if living within a flood zone is driven by economic necessity, the danger or discomfort of living in an area where flooding does happen from time to time can also be resolved by technology; resolved, in turn, by economics. If the cost of redesign is greater than the economic benefit of residing there, then moving somewhere else must be the right answer. Unfortunately, human beings are bad at following this reasoning before they experience flooding at first hand. This may be one reason why floating houses exist as an approach to flood risk mitigation in the Netherlands, but in practice it is hard to find evidence of widespread use!

'Did the Earth move for you Nancy?' is a memorable line from a 1970s pop tune (the somewhat apposite 'Money's Too Tight To Mention', later performed by Simply Red – sadly, one of the authors knows all the words to this). For the most part, human beings prefer the Earth not to move. Nevertheless, many cities are situated in zones known to be regularly subject to significant tremors. Some very large cities, such as Tokyo and San Francisco, sit on top of seismic faults. Yet, this is not a problem if the right technology is used to construct the built environment. High buildings in such cities incorporate far more steel than their equivalents in cities that do not suffer from unstable land. In the Kobe earthquake (the Great Hanshin Earthquake) of 1995, the main injuries and deaths reportedly happened in the context of buildings not built using the principles of modern technology – so, heavy slate roofs sitting on top of a wooden house collapsed on top of those sleeping inside and expressway pillars not built of the correct cement were said to have been the ones that keeled over.

The most devastating quakes, in terms of lives lost, happen when buildings are not designed to withstand shocks. If earthquakes are not capable of moving people away from locations such as Tokyo and San Francisco then something very powerful must be keeping them there and of course this is economics, which, as discussed, ultimately drove the choice between moving to a safer place or designing buildings to make them fit for purpose in an earthquake zone.

Infrastructure is another aspect of location that will determine the economic and financial value of homes, as well as the environmental footprint of their residents. The cost and feasibility of new housing development (therefore an adequate supply of homes) is partly determined by the presence of an adequate infrastructure, but infrastructure is a constrained resource. The consequences of this are demonstrated in the UK, where the location of significant parts of the UK's service sectors in London and the south east has led to a structural shortage of housing – reflected in the land values cited above.

There is something of the chicken and egg about this: the development of a significant transport infrastructure in the south east has historically tended to increase interest in living and working in the south, exacerbating the problem.[3] What happens when infrastructure limits are reached is demonstrated by the experience of people living in the area of Los Angeles, who regularly face a couple of hours sitting on an expressway on the way into work, and on the way back again. There is no alternative mode of transport and so the only choice this consumer has is to find a better-connected place to live. As might be imagined, the cost of doing this is likely to be prohibitive for most people. 'Telecommuting' can work in knowledge-based industries, but not all jobs can be done on that basis. So, all involved are trapped in a resource-intensive, energy-intensive way of living, urging a significant redesign of the built environment, of which housing is a significant component.

As the above illustrates, technology does not always keep up with the problems posed by urban living, and this is often because long-term considerations of 'sustainability' tend not to be incorporated at the design stage, whether because the need is not anticipated, or because the cost of doing so is prohibitive for the supplier.

The twin crunches and housing

To consider how exactly the financial credit crunch will work through the housing environment, let us consider the situation from the perspective of economic first principles. How will the financial credit crunch impact housing supply, and what does that mean for the environment? And how will the financial credit crunch impact demand, and what does that entail in terms of the environmental credit crunch?

Housing supply in the wake of the financial credit crunch

The economics perspective

Housing basically comes in three forms: privately owned, privately rented and social. Economically, the distinction is important when it comes to demand. For supply, the considerations are remarkably similar for all three forms of housing.

Building houses

Home builders (for any market) will suffer two consequences from the financial credit crunch. The first is that, like any business, there is a shift in the cost of funding for home builders. It is more expensive to borrow. Given that housing is now perceived as a more risky business (house prices can go down as well as up – who knew?), there is perhaps an additional risk premium over and above that which is levied on non-construction borrowers.

The second consequence is that housing supply is also affected by the value of land. In a more credit-constrained environment, where house price to income ratios decline, the value of 'land banks' held by construction companies is negatively affected. This is important. Construction companies often use their stock of land as collateral for borrowing. The decline in house prices reduces the value of that collateral, and thus impacts the cost of borrowing, independent of the general funding cost increase arising from the financial credit crunch.

This means that the intrinsic value of new housing stock is likely to suffer. New buildings are going to have to be constructed more cheaply than before, because the costs associated with construction are rising – and rising in a more sustained manner. An extreme example of this, albeit not from the owner-occupier sector, would be the construction of social housing in the UK. The Labour government, elected immediately after the war, started to build houses to a generally high quality. However, large swathes of the population remained in temporary accommodation, particularly in London. The Conservative government of 1951 changed policy. Unable to afford the old housing programme, housing minister Harold Macmillan started the construction of tower blocks.

The housing supply increased, but the quality (both general and environmental) suffered hugely. Environmental problems started almost immediately. The concrete tower blocks suffered from condensation problems. The government allegedly denied that this was a fault of the design or the poor quality of construction. Instead, blame was attached to the (working class) occupants. The working classes, it was held, breathed too heavily and sweated too much – causing the excess moisture and thus the damp problem.[4] This is not, perhaps, an explanation that stands up well to rigorous scientific analysis. The quality of construction seems, perhaps, a more rational reason.

The specifics of the private rented market

Supply of rented accommodation is also directly affected by the financial credit crunch. The 'buy-to-let' market was popular at a time when property prices were rising so rapidly. Here was a chance to acquire an asset that was only ever going to go up in value (sic), and which would pay an income that meant it could be self-financing. Buy-to-let mortgages were typically somewhat stricter as to their credit standards, but during the credit boom this did not place a particularly onerous burden on the borrower.

The provision of credit for rental properties is now considerably more constrained. Although there may be some increase in the supply of properties if principal homes are rented (something we discuss later in this chapter), there is little prospect of an increase in credit for second or third homes being acquired as part of a rental property empire. Where credit is available in this manner, the deposit required is likely to be greater, and the overall cost of debt service (including additional premiums for risk in the interest rate) is also likely to be higher.

The financial credit crunch makes the cost of supplying rental property higher, therefore. This can be recouped through higher rents, or through reducing the quality of the property provided for any given level of rent, or some combination of the two. Rent is partly constrained, of course – housing has to be affordable and the cost of rented accommodation needs to be considered against the cost of alternatives. If a prospective tenant still has the option of obtaining a mortgage, then the relative cost of a possible mortgage will be a limiting factor on the rent.

This then presents a landlord with a problem. The mortgage costs associated with an owner-occupied property are likely to be lower than the mortgage costs associated with a buy-to-let property. Owner-occupiers are generally considered lower risk (the risk associated with them is the risk of job and income security). Buy-to-let mortgages are higher risk (the risks here are those associated with the market for rental properties). Therefore, the cost that needs to be defrayed (the mortgage cost of letting a property) is likely to be higher than the limiting factor on the rental income (the cost of someone being an owner-occupier). It is not a universal problem, and certainly does not apply to low-income housing (where there are different limiting factors on the rent), but it is something that suggests economies will have to be found elsewhere. This then suggests that private landlords have an incentive to reduce the quality of the housing stock that is on offer.

At an extreme, this suggests slum landlordism. This reduces the quality of housing without reducing its price. Less care is taken with the infrastructure; unnecessary (maybe even necessary) repairs are not undertaken. From the landlord's perspective there is no market incentive to improve the environmental efficiency of the property. The landlord pays for the capital cost of updating to a fuel-efficient boiler (for instance) but the tenant is the one who reaps the benefit of the lower fuel costs.

Social housing, the financial credit crunch and its ramifications

Social housing is the final alternative – and its provision varies greatly from society to society. The models used for social housing vary widely, but there are now two broad routes – government financed, and some kind of social housing provision in the private or quasi-private sector.

Government-financed social housing is affected by the latest iteration of the financial credit crisis – the sovereign credit crunch. Governments are running deficits and markets are focusing on gross funding requirements. Far less attention is paid to *why* a deficit is being created. Economists often argue that a government deficit that is used to acquire assets (social infrastructure and so on) should be regarded differently from a deficit that is being used to finance transfer payments. In the current market environment, the niceties of where the money is going are generally being lost in a desire to purge the government balance sheet of unnecessary spending.

In a more constrained borrowing climate, it is inevitable that government plans for social housing are likely to be more constrained. This applies to both the construction of new social housing and to the maintenance of the existing housing stock. The former clearly suffers along with other forms of government spending. The maintenance, meanwhile, is likely to focus on the urgent rather than the important. Urgent maintenance might be repairing a roof or ensuring a safe power supply. Important maintenance might be upgrading the insulation. Most environmental improvements can be seen as important but (considered from the perspective of the market) not urgent.

Of course, some social housing is provided by the private or quasi-private sector. Charities, or (in the UK) housing associations, can play an important role in providing social housing. However, the funding model for such housing is often contingent on the private housing market and, as such, is vulnerable to the financial credit crunch. Many housing associations provide mixed housing developments. The housing association will acquire land and will create a housing development in phases. Essentially, houses are built on the land and sold on the open market. The profit generated from this is then used to pay for the construction of social housing (or mixed-ownership properties) on other parts of the land. The problem with the funding scheme immediately becomes obvious. If house prices for owner-occupied housing are going to be lower (relative to incomes, at least) then the income that a housing association can derive is going to be lower. This means either that the ratio of social housing to owner-occupied housing must shift, or the quality of the newly built social housing must fall, or some combination of the two.

This is particularly important. Social housing is generally targeted at lower-income groups (obviously). These groups are the least likely to be able to afford the capital spending that environmental upgrades normally require. Moreover, there is limited incentive to upgrade a property that one does not own. Therefore, the environmental efficiency of social housing is likely to be dictated in a large part by the initial quality of the environmental infrastructure. If that is constrained by more limited funding under existing finance models, then the environmental efficiency is unlikely to be enhanced subsequently.

The economics of housing supply post the financial credit crunch is fairly clear. There are going to be constraints on supply. Home builders (be they private or public) are likely to emphasize the 'efficiency' of their construction, but they don't really mean 'efficiency'. Generally speaking, the coming era is likely to be marked by 'cheapness' in housing construction. This may not be quite as awful as the housing estates created by Harold Macmillan in the 1950s, but it does not create an auspicious backdrop to the environmental credit crunch. This is the issue we now address: just how bad is a cheaper housing market going to be for the environment?

The environment

The obvious conclusion to drop out of the current situation is that, in financially constrained circumstances, the economics of housing supply will not shift towards an environmentally friendly model. Indeed, the financial credit crunch might induce a less environmentally friendly construction industry. The issue, clearly, is expense. If (in financial terms) the cost of environmental efficiency in construction is higher than less environmentally friendly alternatives, the supplier of housing (be they private or public) is likely to take the less environmentally friendly route. This applies even if the running costs of the home are higher: builders do not have to live in the homes they construct.

Stocks and flows

A home can be described in terms of stocks and flows of energy – stocks embodied in the building materials used to construct the home and flows in the energy used to run the home. Of course, the flows used to run the home will be determined to a large part by the initial infrastructure – how efficient is the house as constructed? As mentioned in the Preface, issues such as heating, water use and other environmental concerns depend on *how* the home is initially built.

Flows and construction

The most significant environmental impact of the residential housing sector arises from energy flows in the form of the energy needed for space and water heating – probably about three-quarters of the total with the rest going to other electrical equipment and gadgets. Since the flows of energy used through the life of the home are far more significant than the stocks of embodied energy, it seems to make more sense to focus on the energy efficiency of the home. To this end, the 'zero-carbon' home is envisaged by the UK government by 2016, but this can apply only to newly built homes. In the meantime, only 1–2 per cent of the housing stock in the UK is renewed per year, implying significant potential energy waste still embedded in the built environment over the medium term. We will come back to the issue of the existing housing stock when we consider the environmental consequences of housing demand.

For the new-build home, initial investment in insulation is clearly going to be important to the long-term environmental impact of the home. However, if the (economic) cost of that insulation is not recouped in a higher sale price, the financial credit crunch suggests that home builders are unlikely to voluntarily pursue insulation. Even if they do insulate, how they insulate still has an environmental consequence. 'Using sustainable insulation materials offers no benefit in terms of thermal performance', or, for that matter, in terms of economics. Sustainable insulation material could cost 50–100 per cent more than non-sustainable materials according to one writer. The same writer also points out, however, that 'the benefits on the environment, the house, and the people that live in it are many' (Pullen, 2008, p51).

In the post-financial credit crunch environment we have a rather chilly outlook for insulation: home builders now have less economic incentive to provide well-insulated homes up front; even if regulation forces insulated home construction, home builders now have less economic incentive to use sustainable insulation materials. The financial credit crunch suggests that housing construction becomes less environmentally friendly – and in fact exacerbates the environmental credit crunch.

The environmental issues of building the housing stock

Even though energy and other resources embodied in the materials used to build homes appear to be a subsidiary issue to energy used by the typical residents, about half of all raw materials taken from the environment are used in buildings. Moreover, about half of all waste production in the UK comes from the building industry. This suggests that how construction is undertaken needs to be considered now so that the development of new building practices and materials innovation (both of which can take time) can be developed.

Box 5.1 Recycling stock and avoiding the flow: The Earthship

An Earthship[5] is a new concept in construction, and yet it is also a very old concept. It is a self-sufficient abode needing no connection to the pipes and wires that supply water and electricity to conventionally developed market homes. It has earthen walls thick enough to act as storage heaters so that the house stays warm at night and cool in the day. Up to this point, this is also a good description of a Gobi Desert farmhouse, but here the resemblance stops: for these UK-based Earthships, 70,000 litres of rainfall per year are 'harvested' from the roof to be stored in underground cisterns. Planters process grey water and reed-beds process wastewater. Wind and sun energy feed a bank of batteries to supply energy needs, and as the house needs very little space heating this is more than enough for the needs of the building each year. This sounds like the pipedream of an environmentalist, but such prototypes exist in Fife and Brighton. The walls are not quite earth: they are built of old tyres, each of which has been tamped full of earth, and once these have been laid in the same way as bricks (1000 of them needed for a home 130m^2), they are clad in tamped earth and then covered with a shell of reclaimed timber and other natural materials.

Building such abodes in scale is, for most governments (or the post financial credit crunch private sector), impossible, in practical terms. Dealing with the housing needs of large numbers of people means large urban centres, and that means building vertically, requiring a more advanced form of technology than earth-filled tyres. Moreover, the house just described would not be everyone's cup of tea. The laws of supply and demand will not, left to themselves, deliver a large fleet of Earthships. Nevertheless, the basic principles – using recycled materials, materials sourced nearby, emitting the minimum of waste to air and water – could be applied far more than they are in housing in the first decades of the 21st century.

In practice, this will be complicated. One commentator, Tim Pullen, has set out a number of sustainable options for sustainably built homes (a few of which are tabulated below). Each of the products named below will be driven by a specific set of economics. Information on the environmental profile of a building, which is not fully delivered in a market context today, will be required. This is not to say the environmental profile of many building materials could not be priced in – in fact, for something like timber, it should theoretically be possible to make use of non-Forest Stewardship Council (FSC) timber (see Chapter 6 for more detail on the FSC) punitively expensive through regulation, and then markets would be able to do their work.

For sustainable housing to become more than a nice idea, someone needs to do the monitoring and measuring in respect of what is used where (along the lines of Table 5.1); this is where organizations such as BRE come in. BRE's

TABLE 5.1 Sustainable building materials: A complex, information-hungry idea

Element	*Preference*	*Options*	*Avoid*
Foundations	Reclaimed aggregate concrete	Concrete and block	–
Intermediate floors	FSC timber joists with reclaimed timber boards	FSC timber	Concrete
Roof structure	FSC timber trusses	Non-sustainable timber	Steel
Gutters, pipes	Galvanized steel	Coated aluminium	PVC, zinc, copper

Source: Pullen (2008, p93).

'Green Guide to Building Specification' rates products from A+ to E on the basis of a range of environmental impacts including embodied energy, based on a lifecycle assessment (LCA). Thinking in terms of the building lifecycle is critically important because the performance of some materials – for instance, insulating materials – can degrade over the lifetime of the home, under the influence of damp, or age, or simply because they were badly installed in the first place. The complexity of the issue may once again present challenges but overall it is possible to envisage a shift in thinking such that dwellings built sustainably are (culturally speaking) valued more highly than those that are not. Of course, economics cannot be ignored, and if this should translate to a price premium in the housing market for sustainably built homes, a problem immediately arises in the context of a credit crunch – namely, that financially constrained consumers may not be able to pay more for sustainability.

As the discerning reader will remember from the chapter on food (Chapter 1), the organic food industry has ceased to expand (indeed, has contracted) as a consequence of the financial credit crunch. So, surely economics spell the death knell for sustainable homes? Well, not necessarily, as the food industry illustrates. What appears to have happened in the food sector (even as the organic category was going into reverse) is an increased adoption of sustainable practices by firms in the 'mainstream' in respect of materials such as palm oil and chocolate. As a consequence, the market in sustainable foodstuffs has actually kept growing and the 'organic' market is now actually the tip of an iceberg (rather than the entire ice cube). A similar development is possible in the housing market because it is driven by a combination of consumer preference and an (increasingly environmental) raft of government regulation, and because if the sometimes maligned corporate sector is good at anything, it is innovating to be able to compete on price at the same time as giving the consumer what he or she wants.

Nevertheless, the problem must not be underestimated and the problem, of course, is that in the wake of the financial credit crunch, and in the absence of any regulatory adjustments to the market price of sustainable versus unsustainable building materials, the economic imperative goes all the other way. The financial credit crunch will make the environmental credit crunch worse, in this regard at least.

Waste not, want not, once again

Governments interested in reducing landfill are also likely to be interested in waste reduction. According to the Chartered Institute of Building, the UK building industry is responsible for 72 million tonnes of waste per year (to be fair, this is more than just new-build, but will include building work on existing housing stock).

This suggests scope for a reduction in raw materials use and other environmental impacts, from recycling as much as possible. What can be recycled is likely to depend on the original design of the building. At one end of the scale, it may be possible to reuse the entire building, simply refurbishing. At the other end of the scale, it may be necessary to exercise care in deconstructing (not 'tearing down'!) for as much of the raw material as possible to be recycled, unless the building was designed for deconstruction (by being constructed in a modular fashion, for instance). So, once again, and, perhaps unsurprisingly, we return to design as fundamental to the sustainability of homes, throughout the entire lifecycle of a building from ground-breaking to demolition.

Waste and recycling are, perhaps, an area where the financial credit crunch may prove to be helpful in tackling the environmental crunch. Dumping building waste in any quantity in the UK now incurs a charge. The growth of architectural salvage (fostered by the planning regulations as well as by market forces) generates a financial reward for recycling. As we saw with food back in Chapter 1, the financial credit crunch does provide some disincentive to waste existing resources. However, careful demolition takes time, which costs money, which is in short supply in an age of austerity. The prospects for managing the environmental credit crunch are more positive in the field of construction recycling than in other aspects of housing, but they are not unambiguous.

Housing supply and the twin credit crunches

For housing supply, therefore, the economics of the financial credit crunch suggest a negative feedback to the environmental credit crunch. Straightforward economics suggests that home builders will seek to produce cheaper homes – be that for private owner occupation, the rented sector or for social housing. This suggests a bias to cheaper materials in construction, which at the moment prejudices against environmentally sustainable inputs. It also suggests that builders will not incur any extra expense to create a more efficient home (in terms of flows such as energy use). As long as builders do not have to live in their own constructions, and while flow costs have little impact on the value of a property, here is little incentive for them to minimize running costs.

If housing supply is a negative story, is housing demand going to prove to be any more optimistic? To consider this, we will break down housing demand into owner occupation (which is generally mortgage financed) and rented accommodation.

Housing demand in general

The demand for housing is driven by a number of factors. Of course, economics has a significant influence – specifically, the level of economic activity, the level of and growth rate in average incomes, consumer confidence, the level of interest rates and the availability of credit.

As discussed in earlier chapters, the two main reasons for the increasing threat of an environmental credit crunch are the expansion in the population and the increasing hunger for resources that tends to accompany economic development. Both factors also happen to be a significant driver of demand patterns in housing. In 1950, the average size of the UK family home was 85m^2; in 2007, it was 130m^2, a rise of more than 50 per cent in a few decades (Pullen, 2008, p15). This shift is for all housing – owner occupied, private rented and social combined.

Not long after the Second World War, lower incomes and perhaps habit (for instance, more people to a room and a smaller number of bathrooms per home) accounted for smaller home sizes. However, the demographic mix has also evolved. The number of single family households in the mix has risen, driven by social factors such as the fragmentation of the nuclear family household, delay in marriage to beyond the early twenties in developed countries, fewer marriages, a higher divorce rate, an increase in life expectancy, improved health and therefore independent living for longer of older generations, and so on. Together this amounts to a growing housing sector environmental footprint in terms of land requirements, materials and 'operating costs', in the form of energy use, water use, and a larger volume of furnishings and other possessions.

In a potential reversal of the trend in the early years of the 21st century, young adults are tending to stay in the family home for longer because homes have become unaffordable, job opportunities more constrained and income growth stagnant at best. Here is a potential indication of the financial credit crunch on general housing demand – a condensation of housing demand into family units that stay together for longer. (This is not a costless option. Condensing housing demand may reduce environmental and potentially financial costs. Sharing a home with one's parents for a longer period of time may increase social costs; for instance, patricide and matricide.)

What this suggests, overall, is that the financial crunch may have scope to shape the environmental footprint of the housing sector, if only by reversing (or perhaps just stabilizing) the post-war trend demand in developed economies for more and more space per person.

Long-run trends in demand (for owner-occupied housing, but also for rented housing) are also driven by things that directly affect the quality of life in any given residential dwelling, such as its connectedness to a functioning infrastructure (transport, energy, water provision, waste collection) and the quality of the existing housing stock. One of the arguments surrounding the social legacy of the 2012 London Olympics, for instance, is that the quality of life in East London will improve as a result of improved transport and social infrastructure. A key measurement metric for how the quality of life has improved is, in fact, any change in relative house prices in the area.[6]

The economic and environmental impact of housing is not just about overall demand for housing stock (demographics and condensation). It is also about what sorts of housing stock are demanded – there are distinct consequences from the choice between owner occupation and renting.

Housing demand in the wake of the financial credit crisis: Owner occupiers

The economics

Levels of owner occupation vary significantly from country to country. They also vary across time. The UK is regarded as a society of home owners – 'an Englishman's home is his castle' is seemingly one of the defining features of the national character. Today, not far off 80 per cent of the UK population own their own home. The concept of a nation of owner occupiers is, however, an essentially modern myth. In 1900, only around 10 per cent of houses were owner occupied. The vast majority of the country, including the middle classes, lived in rented accommodation (Smith and Searle, 2010, p16).

Similarly, the US perceives owner occupation as wrapped up in the 'American Dream'. The white picket fence, the flag on the porch – this is an important aspiration for the aspirational American citizen. The boom that preceded the financial credit crunch led to a general increase in owner occupation. The credit expansion was specifically aimed at increasing the scope of credit provision. People who had previously been allowed nowhere near a mortgage were handed forms, told to indicate their income levels and reassured that no one would do anything so crass as to check the veracity of the statements made. In the US, increased home ownership was seen as a desirable political objective. Pressure groups formed to hound banks into lending to lower-income groups and to loosen credit standards in order to achieve the greater home ownership. This is not to say that banks were helpless victims of manipulative politicians and consumer groups – banks welcomed and embraced the chance to take on more mortgages.

One reason that banks were willing to take on higher amounts of mortgage debt (and to extend mortgages to a wider spectrum of society) was the fact that mortgage debt could be securitized. In this way, the risk associated with mortgages to lower-income groups could be passed off to investors – or so the theory went. As a result, mortgage lending exploded in several key economies. Accompanying this, the instances of home ownership increased. In the US, home ownership went from 64 per cent to 68 per cent of the population over the period 1990–2007.

Of course, the increase in home ownership was lower than the increase in mortgage debt. This is because of mortgage refinancing ('having a cash machine on the side on one's home'), which allowed consumers to borrow against the capital value of one's home in order to afford life's little essentials (a fifth sports utility vehicle, or a 52-inch plasma television, for example – without the possession of which, human existence can barely be tolerated).[7] There was also an increase in second-home ownership, which was financed either with a second mortgage or with an extended mortgage on the consumer's principle residence.

So what does the financial credit crunch do to owner occupation? Essentially, there are three consequences. First, the cost of owner occupation is likely to rise. Second, the universe of people able to participate in owner occupation is likely to shrink. Third, the price of houses relative to income is likely to come down.

The cost of owner occupation is likely to rise because, of course, the cost of having a mortgage has risen as a consequence of the crisis. This does not necessarily apply to existing mortgage holders (who may have locked into interest rates in the boom of easy money). However, new mortgage applicants are likely to face two additional costs. Obviously, the risk associated with lending has risen. Increased regulation, restrictions on the use of securitization to defray risk and (frankly) the return of common sense to lenders is raising the risk premium associated with mortgage lending. Higher risk means higher interest rates to compensate lenders.

The second increased cost is that lenders are starting to require higher deposits. This is a less obvious cost, but it is nevertheless something that economists pay a lot of attention to. If one is allowed to have a mortgage at 100 per cent or even 110 per cent of the value of the property, there is virtually no obstacle to purchasing a home. If one has to save up 5 per cent or even so unimaginably high a sum as 10 per cent of the value of a property as a deposit, it imposes a cost on the consumer. (In some parts of continental Europe, deposits of 20 per cent or 30 per cent of the value of a property are required. This is obviously beyond the comprehension of an Anglo-Saxon audience.) Consumers have to forgo other consumption and build up idle cash balances to fund this deposit. That is an expense that may deter some potential homeowners.

The universe of potential homeowners will shrink with the financial credit crunch. This is partly because the increased cost will put home ownership out of the reach of some of the lower-income groups in society. Price up, demand down is one of the basic rules of economics. However, tightening credit standards also suggest a less sophisticated credit allocation. In the wake of the financial credit crunch, mortgage providers are likely to issue blanket bans on certain categories of potential borrowers. Thus, even if a consumer could potentially afford a mortgage, the lenders will not offer credit because the category into which that borrower falls is considered too high a risk. This lack of discrimination within categories of society was attacked by consumer and political groups during the credit boom. The fact that these groups spent the boom years urging lenders to lend more, effectively neuters their influence now (even if some discernment in lending is actually desirable).

Finally, house price to income ratios are likely to be affected by the financial credit crunch. Demand for housing is being reduced, which (all things being equal) will reduce the price of housing. In particular, the amount of credit that lenders will lend is now likely to be reduced (to lessen risk). House prices will fall as a consequence. The relationship between house prices and the amount of credit available is pretty intuitive. If the average mortgage is set at five times the average income, then the average home buyer has the potential to spend up to five times their income on a house. The mortgage money is something that the consumer will spend. They may start by being restrained, and only prepared to buy a house at four times their income; but if someone else outbids them on their dream home, the home buyer will raise their offer up to the limit of their resources. The limit of their resources is five times their income, so eventually that is what they will bid. As housing is an emotional purchase (wrapped up in the concept of middle class respectability, the American dream, and so forth), a calm and rational assessment of what *should* be paid for a property is subsumed into a desire to pay whatever can be afforded.

In the wake of the financial credit crunch, mortgage lenders have reduced the leverage that they are prepared to offer. Independent of other changes in the demand pattern, this will reduce average house price to income ratios. This means absolute house price declines in some regions, or static house prices in others (as the adjustment in the house price to income ratio is driven by rising incomes). So, the financial credit crunch means owner occupation is more expensive and fewer people are allowed the opportunity to purchase a home. It also means that house prices are likely to fall, at least relative to incomes.

This last point may be confusing. It seems that we are arguing that housing is both more expensive and cheaper at the same time. In fact this is not so. The price of the house may be lower, but the price of borrowing the money necessary to purchase a house is going up. If one is a cash buyer, the price of housing is in fact coming down (as there is no borrowing cost). For the majority of the population, however, the cost of buying a house is rising, even though the value of the asset purchased is coming down.

The demand for owner occupation is likely to fall in the wake of the financial credit crunch. Fewer people in the future will seek to own their own home or be able to own their own home. Already owner occupation numbers seem to be slipping in both Europe and the US.

The environment

What does the reduction of owner occupation mean for the environmental credit crunch? There are two opposing forces.

Reduced home ownership means, in fact, that less money may be spent on maintaining property in the country as a whole. Home improvement spending is likely to be most concentrated on one's own home, where one can enjoy the benefits. If you own your own home, you pay the running costs on that home – the capital spending for a new boiler, insulation or a new conservatory. Capital spending for environmental and economic efficiency is likely to decline.

Alongside the reduction in home ownership, however, we are also witnessing a decline in housing market transactions – or housing churn. With increased costs associated with housing, and borrowing credit becoming more difficult, people may be reluctant to move house. This is particularly evident in the UK where the number of housing *transactions* has declined by almost 50 per cent from 2007 to 2009 (1.62 million transactions in 2007 fell to 0.85 million transactions in 2009). If, by moving house, one has to renegotiate a mortgage (on less favourable terms), the cost of moving house rises. An owner of a pre-crisis mortgage is unlikely to get those terms again, in the chastened and conservative credit environment we now find ourselves in.

This then creates a countervailing force. If an existing homeowner, with a pre-crisis mortgage, will not move house because of increased borrowing costs, then they will contemplate improving their existing home. This is something we examine in more detail in the next chapter on consumer durables. In the housing context, this may mean an increase in energy efficiency in the house (through installing a new kitchen, for instance). Energy efficiency may not be the primary objective of the improvements, of course, but it is a consequence nonetheless. It may also increase housing density. If one cannot afford to move to a larger house but can add an extension to one's current home, then more people can (potentially) occupy the same land surface.

The environmental consequences of owner occupation demand shifts basically come down to whether the existing housing stock (in total) becomes more or less environmentally efficient. This is about retrofitting.

The energy profile of a building depends on how energy efficient it is – a combination of the installed heating system and how well insulated the building happens to be. Insulation materials have the greatest potential to improve the energy efficiency of existing homes through retrofitting programmes. Thermal imaging technology has been used to demonstrate how energy-inefficient the UK's housing stock is, suggesting that significant energy savings could be made by a country-wide initiative to insulate homes properly. A desire for energy efficiency (covered in Chapter 3) and a desire to improve one's current home (because of reduced housing churn) could create a positive environmental consequence. Indeed, government grants for insulation are in place and the energy services provided by some utility companies extend to home insulation. The question is whether such measures can survive in the age of austerity; they are a critical economic incentive to exploit the shifts in consumption that the financial credit crunch has generated.

The water efficiency of a building depends on the water infrastructure in the house. Very few homes have the capacity to recycle grey water or to harvest rainwater because the home is not furnished with the necessary equipment. Water-efficient washing and washing-up machines, taps, showerheads and toilets are on the market. If housing market churn encourages home improvement, then the improvement can be more environmentally efficient.

As owner occupation falls, the retrofitting of that proportion of the housing stock that has ceased to be owner occupied is unlikely. However, the remainder of the housing stock has potentially a greater chance of retrofitting. The problem is that this depends on the consumers' willingness to spend money – and as we saw in the last chapter, this is shaped by the financial credit crunch. Credit, after all, helps spend on capital that provides a benefit over time. The wood-burning stove introduced in our chapter on energy (Chapter 3) is something that improves the housing infrastructure for owner-occupied housing, but that is also under threat.

Government incentives to help homeowners upgrade their homes through retrofitting are also under threat. Several locations have grants for the installation of renewable technology, but the survival of these schemes has to be considered questionable as governments seek to cut spending. For instance, a pilot 'Pay as You Save' scheme (described on the Department for Environment, Food and Rural Affairs' (Defra) website) circumvents constrained credit markets by offering residents of the London Borough of Sutton (government) loans of up to £10,000 to pay for a mix of energy efficiency and renewable energy measures (boiler upgrades, cavity wall insulation and solar panels). Repayments are set over 10–25 years so that 'the amounts householders are paying each month should be lower than the amount they are saving on their energy bills'. The government is acting as a bank providing credit – because of course the banking sector is (partly at government instigation) becoming more conservative in its lending practices. The problem is that the government is less likely to continue such practices at a time of spending cuts, and is certainly unlikely to extend the scheme from local pilot to national standard.

Housing demand in the wake of the financial credit crisis: The rented sector

The economics

If owner occupation is more expensive, or more difficult, the choices available are limited. Typically, the private rented sector is the next alternative to owning one's own home. In the UK, private rented accommodation accounts for 14.7 per cent of dwellings. In Switzerland, market rate rentals stand at an astonishingly high 50 per cent of the population. The financial credit crunch does not impact the demand for rented accommodation in a direct sense –

because, of course, one does not have to borrow (typically) in order to rent. However, private rentals are impacted as a residual effect of the credit crunch on the owner-occupied sector. Aside from the obvious binary choice (own or rent), it is possible that both options are chosen simultaneously. There is evidence that people own (though do not occupy) a property while renting a home as a consequence of the credit crunch.

There are two routes to the simultaneous 'owning and renting'. The first is downsizing, and is a feature of the US among other economies. If a home-owner cannot afford mortgage payments on the family home, one option is to move out and rent a smaller house. The owned property can then, itself, be rented. The idea, of course, is that the rental income will cover the mortgage payments on the original property, while the rental outgoings on the new property will be more affordable. There is no need to realize a capital loss on the original property if property prices have fallen, and there is always the hope that better times will return and one will be able to move back into the original family home. A variant on this is more prevalent in the UK, where mortgages are associated with both the individual and the property. As moving home will necessitate repaying the original mortgage and taking out a new mortgage (almost certainly on less favourable terms), there is a strong desire not to move – as we saw in the preceding section on housing churn. However, changing circumstances may force someone to move house (job relocation, for instance). If the cost of changing mortgage is too high, consumers may prefer to hold onto their original home – renting it out – while moving into a rented house in the new location.

Social housing in the wake of the financial credit crunch

In extremis, there may also be an increase in demand for social housing as a result of the financial credit crunch. Social housing could be an alternative for people who would previously have sought to own their own home (a reversal of the trend from the 1990s, in effect). Again, this is not a pull phenomenon (people choosing social housing), but a push phenomenon (people not able to choose owner occupation).

The environment

Earlier, we divided housing into the concept of stock (building the house) and flow (the consequences of occupying the house) – and it is obvious that it is flow that matters most in environmental terms. A variation on that theme is the stock of building a house and the flow of maintaining it. The motives that surround maintenance in the rental sector – private or public – are very different from those in the owner-occupier sector.

We saw with owner occupation that there may be a boost to retrofitting houses – upgrading the existing home and in the process (whether or not it is intentional) making the house more environmentally friendly. The benefits of any improvements are felt by the owner directly. The problem with the rented sector is that the owner will only directly benefit if they can charge an increased rent as a result of the improvement. As rent levels are constrained by a myriad of forces, this suggests that environmental retrofitting is less likely.

This may also impact maintenance. Maintenance of a property is generally a case of 'pay now, to avoid paying more later.' For instance, one may pay to paint wooden window frames this year, or pay to replace rotten wooden window frames at some point in the future. Governments in particular have an incentive to delay maintenance of social housing now (because the 'value' of government spending has risen, as the demands of fiscal austerity mean that there is less government money to go around). A poorly maintained home runs the risk of greater environmental inefficiency and a greater economic cost to rectify that inefficiency in the long term.

What else has to happen?

The financial credit crunch is not going to remove the environmental credit crunch in the field of housing. Indeed, the evidence is that, on balance, the environmental credit crunch will be made worse. So, what else can be done to push Earth Overshoot Day towards 31 December?

Government regulation

At the level of government, there is increasing awareness over the need to take resource use into account in the design of the typical home; in the UK, 2006 saw the launch of the Code for Sustainable Homes. The code is voluntary and covers a range of design categories: energy and carbon dioxide, water, materials, surface water run-off, waste, pollution, health and well-being, management and ecology. Under this code it will be possible to rate homes for their overall sustainability profile. This code is intended to complement the system of Energy Performance Certificates (EPCs) introduced in June 2007 under the Energy Performance of Buildings Directive. From this date onwards, anyone buying or selling a home is required to provide an EPC, which provides a rating running from A (the best) to G (the worst).[8] The average in the UK is currently D, suggesting that there is considerable room for improvement. This could, if things are implemented in the right way, turn into an opportunity rather than a threat in economic terms.

As the above suggests, a substantial proportion of developing building regulation is being driven by climate change considerations. As discussed, the UK government is aiming for all new homes to be zero-carbon by 2016. A zero-carbon home is defined as one that, in net terms, emits no carbon dioxide from 'operations'; that is, from heating, cooking and other appliances, TVs and computers, and any other use of energy in the home. This definition of zero-carbon could be considered to be too narrow – it fails to capture energy applied in the delivery of water to the home (where relevant) as well as in waste treatment, energy used in materials used to build the home (as well as the freight used to get them there), and it also fails to take into account energy used in travel by the residents. (This last point may sound irrelevant; however, energy used in transport is a direct result of the combination of housing location and work location.) Nevertheless, it may make sense to use regulation focused on 'zero-carbon homes' to focus on the operating energy of the typical household, resolving other issues through other regulation applied elsewhere. As discussed in other chapters, the problem facing many consumers is that they have no choice over the way they use energy because this is embedded in the design of their dwellings, in many aspects of the built infrastructure, and even in the way their jobs work.

The advantage for regulation from the government perspective is that it is generally low cost to the government. The cost is born by the housing provider – and ultimately paid for by the consumer. The consumer may, of course, object.

Innovation

Governments are not automatically thought of as a source of innovation, but they can be.[9] The idea of the zero-carbon home has already sparked some interesting innovations or, perhaps more accurately, departures from normal practice in both the public and the private sector. Unlike the UK, the structure of the German housing market is to operate in groups or cooperatives, and this makes the installation of alternative energy logistically easier when it covers local groupings of dwellings. (Co-housing communities in which heating infrastructure is shared are also well-established in countries such as Denmark and Iceland.)

The Olympic Park for the 2012 Olympic and Paralympic Games is unusual in UK terms in being an example of 'co-housing'. It is intended to be a blueprint for 'sustainable living', the innovation being in the specific combination of solutions. A combined heat and power (CHP) plant is expected to supply the park with energy 'resulting in a 20–25 per cent reduction in carbon emissions in the longer term.'[10] Water consumption is expected to be 20 per cent below the London average. The aim is to recycle 90 per cent of demolition waste. In short, the Olympic Park is a design prototype.

Several governments are flirting with the idea of town-sized 'pilots'. One example is the 400-acre outcrop called 'Treasure Island' on the coast of San Francisco.[11] The plan is to turn it into an environmentally sustainable community for 24,000 residents (the project specifics are private sector; the strategy is government-led). In the village of Stawell near Somerset, five new homes, built on brownfield land, feature on the Defra website as models of sustainability. Energy is supplied through micro-generation, a combination of solar thermal and solar photovoltaic, as well as carbon-neutral wood pellet burners. The houses incorporate rainwater-harvesting systems and efficient plumbing. All timber came from sustainable resources and PVC, solvents, glues and chemical-based paints are avoided. Elsewhere, in Hackbridge, South London, the idea of the BedZED[12] was to deliver a mixed development of 82 homes requiring no fossil fuel to run them. Some of the building design features include very old ideas such as passive stake ventilation, concrete walls that act as storage heaters and a CHP plant. One interesting discovery was that the residents had to unlearn old habits in order to use the houses as designed (for example, not interfering in the automatic heating and cooling process) and also that design tweaks had to be made to the homes (as flaws in the system were revealed), so that the homes would function as expected.

All of the above is encouraging, but will only turn into having a real impact when such ideas become more than pilot schemes, noting that the shift to the 'big society' away from 'big government' in the UK (and constraints on government budgets elsewhere) must be recognized as a potential risk to the continuation of some of them. From a market perspective, when homes are built in this way from scratch, it may be easier for the firm that built them to capture the value of environmental enhancements in the selling price. When it comes to retrofitting, it is likely to be easier for the homeowner to undertake energy efficiency related improvements. From the perspective of landlords, the difficulty of earning a return on investment made to render the home more energy or resource efficient is likely to be a powerful incentive to do nothing about it. Given the likelihood of a post-credit crunch shift away from owner occupancy to rental housing in economies such as the US and the UK, there is clearly a significant risk that the housing-related environmental footprint keeps growing.

As the above makes clear, although a great deal could be achieved at the level of individual homes, efforts to render housing more sustainable at the level of entire communities is more likely to have an impact in moving things forward. What is needed is a new approach to design in housing, and because this means design in the most holistic possible sense, government policy needs to be shaped to allow this to happen. Only when this is in place will economic forces be able to propel things in the right direction from an environmental perspective.

Conclusions

Housing was the canary in the coal mine of the financial credit crunch. It keeled over and died some two to three years ago. However, the economic implications from this are still coming through. As the crisis has spread, to include even sovereign credit today, most aspects of housing have been impacted. Economically, the patterns of housing demand have shifted. This is more than a straight reversal of the housing phenomenon of the past 15–20 years. There is a more profound structural change coming through. Owner occupation is likely to decline over time, which potentially reduces the pace with which housing infrastructure is updated. At the same time, those who enjoy the legacy of a pre-crisis mortgage are likely to be disinclined to move property. That suggests improvements will continue to be made to housing infrastructure. Of course, the proportion with pre-crisis mortgage deals (and thus extra inertia) will be a dwindling proportion of the homeowner pool over time – and their beneficial impact (in terms of infrastructure improvements) will be overtaken by the decline in home ownership. The rental sector – both private and public – is also likely to see less attention paid to infrastructure. With capital gains providing less of the return on private sector rental property, the asset must be 'sweated' harder. That is to say, the rental income net of maintenance costs will have to provide a higher proportion of the overall return on the asset. If the rent cannot go up, because of rent control or market circumstances, then the maintenance costs will have to come down. In the absence of regulation insisting on certain minimal standards, environmental efficiency is likely to suffer.

Housing is unlikely to see a reversal of environmental efficiency as a consequence of the financial credit crunch, even if it reverses demand in developed countries for more space per person, as suggested here. However, advancing environmental efficiency requires technology development, investment, forward planning and sustainable design. These do not come free; therefore, as things stand, environmental efficiency is likely to be considerably more difficult to achieve. It is, however, worth remembering that the twin crunches – the actual credit crunch and the developing environmental crunch – are not the only ones. An energy crunch would go a long way towards prompting greater energy efficiency and therefore could act as a significant catalyst in the right direction. The positive note in this chapter is that there is plenty that could be done, given political will in the right place, and an energy crunch could just put it there.

Notes

1. Earthships are described at the following sources: Hewitt (2008), www.earthship.org, www.sci-scotland.org.uk, www.lowcarbon.co.uk.
2. The powers that be are, however, aware of the issue and consultations are underway. The 'Thames Estuary 2100 Project (TE2100)' is tasked with protecting London and the people living in the Thames Estuary from flooding now and into the next century. The relevant report can be found at http://publications.environment-agency.gov.uk.
3. This process can work both ways. The East End of London traditionally had very short-term property leases (much of the area was originally owned by the Catholic Church, which favoured short-term leases). As a result, there was little investment in the properties, which meant that the area was run-down. This in turn meant that

better-quality businesses shunned the neighbourhood, even when longer-term leases were available, and a cycle of poverty became self-perpetuating.

4 The source of this observation was a *Senscot Public Lecture* marking the 80th Birthday of Professor David Donnison, 26 October 2005.

5 See Note 1.

6 It seems a little bleak to reduce the social benefits of the Olympics to the value of the housing market, but this is (economically) the right assessment. The price one pays for a house is a reflection of the house structure, combined with the environment in which it sits. People will often pay a premium for a house that is near a good school, for instance, and estate agents often advertise (as an important selling feature) the proximity of a home to transport links. If, as a result of the Olympics, house prices in East London rise relative to equivalent house prices elsewhere in the capital, this can be taken as an indication of a legacy effect on the social infrastructure.

7 It should be noted that not all of the expansion of credit in the past two decades was wrong. There were structural changes in the OECD economies (including, for instance, the decline in inflation and inflation uncertainty), which should enable more people to borrow. Securitization, properly handled, is an acceptable mechanism for risk management. The problem is, of course, that things were taken to excess.

8 Ratings information is found at http://epc.direct.gov.uk/.

9 In extremis, governments do innovate. If one thinks of the advances that come about through war, for instance – the Second World War spawned a series of technological developments, not least the computer. The Vietnam War led not just to technical innovation but to innovation in business structures – the pattern of the US airline industry owes much to the organizational innovations spawned in the conflict.

10 Description drawn from Defra's website, http://engage.defra.gov.uk/.

11 See www.oewd.org/development-projects-treasure-island.aspx.

12 See www.peabody.org.uk/media-centre/case-studies/bedZED.aspx.

References

Hewitt, M. (2008) 'Earthships', *Green Building Bible*, vol 1, pp266–271

Pullen, T. (2008) *Simply Sustainable Homes*, Ovolo Publishing, Huntingdon

Smith, S. and Searle, B. (eds) (2010) *The Economics of Housing*, Wiley-Blackwell, Oxford

UK Valuation Office Agency (2010) *The Agricultural Land and Property Market* and *The Residential Building Land Market*, www.voa.gov.uk/publications/property_market_report/pmr-jan-2010/index.htm.

CHAPTER 6

Consumer Durables: Shop till You Drop

Thus every part was full of vice; Yet the whole mass, a paradise. (Mandeville, 1714, *The Fable of the Bees*, explaining why lust and envy for possessions was a collective good for the economy)

The rise and rise of consumerism – and the increasingly ephemeral 'durable good'

It may seem foolish to start with so basic a question, but what is a consumer durable good? There is a hint in the name of course; clearly, a consumer durable good is a good (not a service) purchased by consumers (not companies) that has a relatively long life – two or three years minimum. But what is that? The answer is not static, either over time or when comparing different income groups. In general terms, history suggests that people used to use things for much longer than we do today, even passing them onto descendants rather than throwing them away.

In Stratford-upon-Avon, for instance, Shakespeare's house contains some beautiful objects that illustrate the consumer habits of the day. In his will, William Shakespeare bequeathed to Anne Hathaway his 'second-best bed'. In modest dwellings, such as that of the Shakespeare family, the *best* bed was often on display in the living room, as a piece of conspicuous consumption. Prior to the Industrial Revolution, furniture was made to last and, historically, was a major item. In the late 18th century, a member of the Northern gentry, Elizabeth Shackleton, could record her son's purchase of furniture 'with all the fanfare of a rite of passage' (Vickery, 1998, p167). A Gillow's (thus provincial) mahogany dining table in 1779 cost £5 5s, the equivalent of around £680 today. Of course, in 1779, so monstrous a sum of £5 5s was quite beyond the means of most of the population. It represented a quarter of the annual income of an agricultural labourer who earned little more than £21 in a year, and even 3 per cent of the average income of a clergyman (earning roughly £183).

In 2010, IKEA retailed a dining room table (the Ingo) at £29 – equivalent to 0.1 per cent of the average household's annual income. The quality of the 'Ingo' table may differ somewhat from the quality of the Gillow's table, of course – but that is the point. All but the lowest-income consumer in the UK is likely to regard a £29 table as practically a disposable item and not a durable item. At that price, whether an Ingo table provides service for two or three years is determined by the quality of the table, the dictates of interior design trends and the whim of the purchaser.

Prior to the 18th century, in England, nearly all dinner plates were very clearly durable goods. Plates were wooden or pewter for the most part. Barring fire, or perhaps woodworm, they were pretty much indestructible. Then the concept of 'breakability' was introduced. China (or earthenware) crockery replaced the durable goods of old. Fragile objects sometimes gain in value through rarity value. At the other end of the scale, breakability was accompanied by fashion – and consumers started voluntarily giving up their dinner services (or at the very least, their tea services) because they were no longer in the latest style. By the 20th century, plates can be paper and designed for a single use (many paper plates fail to survive even that).

The rural gentry of Georgian England would even regard clothing as durable consumer goods. Dresses, when they were unfashionable or faded, would be unpicked. The composite lengths of cloth would be sent off to be re-dyed, and then returned to be reconstructed (or made into new styles). In Victorian England, clothes would be used for years: the poorer districts of London recycled clothing (even using it as collateral for loans at local pawnbrokers). It is unlikely that an 'Easycare' shirt from the retailer Matalan (price £5) will ever be accepted as collateral for a loan today; even at the height of self-certifying mortgages, the financial system was more choosey in its choice of security than that. Modern (high-income) economies have a far more disposable approach to clothes – shoes that cannot be readily repaired when worn out; clothes of a fabric that will not endure too many washes. Today, clothing and footwear are not ordinarily classified as consumer durable goods.

The issue of clothing durability was somewhat cynically analysed in the 1951 Ealing comedy *The Man in the White Suit*. Alec Guinness played the inventor, Sidney Stratton, who invented a dirt-repelling and indestructible cloth. In the film, a combination of business-owners and trade unions suppressed this – the ultimate in durable clothing – because were it to be manufactured, the textile industry (and the economic activity that it generates) would be destroyed.

So where does that leave the hapless economist trying to determine what is a consumer durable good? The best approach is probably to be flexible as to the boundaries. A consumer durable good should be considered to be high value (relative to income) and to be reusable (that is to say, have a life beyond its immediate consumption). It should therefore have a relatively long life (three years is the accepted norm) and perhaps (though this is not absolutely necessary) a resale value. The most obvious example of a consumer durable good is a car. It takes a total of 17 years for consumers to have scrapped even half of all the cars produced in a single year (Larsson, 2009). A car is expensive relative to income. A car even has a resale value. This definition of what constitutes a consumer durable good is not perfect, but it clearly classifies a washing machine as a consumer durable good. Food is clearly not. How the Matalan 'Easycare' shirt is to be classified is left to the reader's sage judgement.

Consumer durable goods and the environment in the disposable age

So, having determined, at least roughly, what a consumer durable good actually is, what has been the consumption pattern of these objects in recent times?

Karl Marx, who as well as being a political philosopher can justly lay claim to the superior title of economist, infamously described religion as the opiate of the people. In modern, developed economies, the drug of choice appears to have changed; shopping 'until you drop' appears to be a critical part of keeping voters-at-large content with their lot. One can take the analogy quite a long way. Shopping malls have the scale and perhaps even some of the grandeur of medieval cathedrals. Pilgrimages to such places take place at around carefully ordained points in the calendar. The 'high priests and priestesses' of fashion order regular attendance – and to fail to attend is to become déclassé and thus excommunicated from respectable society.

Improved affordability in consumer durables over the medium term has boosted demand for 'stuff' to fill their homes, making durables accessible to more people and thereby appearing to improve the quality of life for large numbers. Of course, this increase in consumption also has environmental consequences, and it is that interaction that is the focus for this chapter.

Politicians will be sensitive to this. Consumers are happy when consuming. A happy consumer is a happy voter, and that is what politicians care about. If a fall-off in incomes reverses this consumption–happiness relationship (however illusionary it is), then the positive environmental impacts of a return to an old-fashioned definition of the consumer durable could be accompanied by unwanted social consequences should voters become angry about an apparent (even if illusory) fall in their quality of life relative to that of the wealthy. This is something governments will have to deal with sooner rather than later if the impacts of the financial credit crunch turn out to be protracted. However, ignoring the environmental consequences of aggressive consumption may just create bigger problems in the future. The economic effects of the environmental credit crunch may be broader and longer-lasting than those of the financial credit crunch.

The typical 21st century household is a veritable cornucopia of consumer durables – telephone, microwave oven, CD player, mobile phone, satellite receiver, home computer, tumble dryer, internet connection and dishwasher.[1] Even 20 or 30 years ago, this list of possessions would have represented significant wealth. Now they have become commonplace. Moreover, they appear to be following the path of becoming less and less durable (or more and more replaceable), following fashion or new technology.

Looking back to a starting point for this bonanza of consumerism, the 1851 Great Exhibition seems as good a place as any to start. Visitors came to see a celebration of progress, in the form of objects of practical use in everyday life, or objects of beauty, or, simply, objects made in new and curious ways. The best of British was on display along with French silks and fabrics, German musical instruments, toys and furniture, and American agricultural implements. The sheer numbers that visited this exhibition – and the long, hard journeys people were prepared to make – can be read as latent desire for consumer durables. Even so, demand was slow to take off in some countries. Alison Light comments (exploring the lot of the servant in Virginia Woolf's lifetime): 'Vacuum cleaners, electric cookers and water heaters were beyond most pockets; washing machines and fridges an unthought-of luxury' (Light, 2007, p181). Even by 1945, only 20 per cent of British homes had an electric cooker, 15 per cent had a water-heater, 4 per cent a washing machine and a mere 2 per cent a fridge (Light, 2007).

Unlike the pre-Industrial Revolution consumer durables, modern-day consumer durables such as cars, white goods (washing machines, fridges and other household equipment) and furniture seem to be made with the intention that they are replaced after a relatively short period of time – in short, they suffer from 'designed obsolescence'. For household equipment, replacement is likely to be triggered by a mechanical problem, when the tendency is to replace rather than mend, frequently because mending often costs more in both time and money than buying a new one. The impediment to repair often arises because there is a shortage of people able to repair things. The manufacturers of white goods often guarantee their goods for a (relatively short) period of time. After that, manufacturers tend to offer a maintenance contract – at significant cost. The back of the envelope calculation for most is that it will take little time for the insurance premiums to cost the same as replacing the entire machine.

The throwaway culture and the environment

The whole idea of a product being 'disposable' is relatively new. Perhaps the first disposable product was the smokers' clay pipe of the early 18th century. Prior to that, things were meant to be truly durable and be used time and time again. The long-stemmed clay pipes were mass produced, cheap, easily breakable, and in every sense 'throwaway'. The disposable society perhaps dates from this point. In contrast, modern society places such a premium on consumer spending that tax regimes occasionally encourage a short replacement cycle for big-ticket items. In the UK, the arrival of the first MOT test on a new car (on the third year of ownership) is often taken as an excuse to purchase a newer model.

By now, it is clear that the reasons for the 'buy today, throw away tomorrow' consumer culture we live in are complex. From the perspective of the environment, the issue is a misalignment of incentives. It is likely that three primary drivers have been responsible for this state of affairs.

First, the easy abundance of energy and other materials was one thing that made it possible to keep coming up with the 'latest thing' to tempt the pounds from the consumer's purse.

Second, technology innovations have also been important: Moore's Law has allowed more and more to be stored on a chip. (Moore's Law states that technology upgrades should allow computer capacity to double every two years.) This has allowed the functionality of many consumer durables to be transformed – a form of price deflation, because more and more technology is delivered for a similar (or lower) price. Battery technology has allowed a process of miniaturization, making it extremely attractive for consumers to upgrade computers and phones at regular intervals, and car manufacturers spend billions of dollars on research and development (R&D) to come up with the latest must-have engine. Put together, these forces encourage a 'throw away and replace' mentality.

Third, the rapid development of capital markets and credit markets has had two important effects: it has stimulated the growth of the large firm competing on the ability to grow market share, and this is often done on the basis of giving consumers an incentive to upgrade existing goods, by tapping into the urge to 'keep up with the Joneses'. It has also allowed finance to flow to the consumer: it is no coincidence that large auto firms often also have a significant finance arm.

The economics of the environment and the economy

As consumers are increasingly aware, the relationship between durable goods and the environment is a two-way process. Consumption of anything involves an environmental impact, and durable goods are no exception. The frequency with which durable goods are purchased and the manner of their eventual disposal (as well as whether they are disposed of at the end of their lives, or midway though) all has a bearing on the environment. In this sense, the environment will benefit if the purchase of consumer durable goods is kept to a minimum.

On the other hand, many consumer durable goods have environmental impacts through their operation. A car, clearly, consumes fuel. A new car generally consumes less fuel than an old car (for the purposes of this argument we chose to ignore some of the more monstrous SUVs seen in the 21st century high street). A washing machine consumes fuel and water – and the newer the washing machine, the more energy and water efficient it is likely to be. The important point here is that, because a consumer durable good provides an economic service for several years, and because all economic activity has an environmental consequence, the environmental implications of purchasing a consumer durable good are long-lasting.

Indeed, some economic commentators consider durable goods a form of saving – acquiring an asset in order to benefit from an economic service in the future. Conventional saving, for example opening a bank deposit account, is acquiring an asset in order that one may purchase goods in the future (and thus derive economic benefit – economic utility as it is somewhat dispiritingly known). A durable good, stretching definitions a bit, can be thought of in the same terms. One does not buy a car because of the car per se – one buys a car in order to enjoy the economic benefit of transport in the future.

All consumer durables have an environmental footprint – resources abstracted to make them, other resources used in the making of them and by-products such as waste left behind in the environment after manufacture, and again once they have reached the end of their useful lives. Clearly, modern behaviour (and modern design) means that consumer durables are readily and regularly switched for a more up-to-date version, these objects are better described as 'consumables' rather than 'durables', and therein lies a significant environmental problem. A burgeoning stream of disposable consumer durables means a burgeoning stream of wasted resources. The environmental impact of consumer durables could thus be reduced in three ways: by designing them so that fewer resources (such as water and energy) are required to use them; by using them as their names suggest they should be used, for long periods of time; or by recycling them or their component parts.

The desire to purchase environmentally friendly consumer goods

We began by suggesting that the 'desire to have' the relevant consumer good together with price, income and credit combine to drive purchasing decisions when it comes to consumer goods. Matching the bundle of services or other benefits delivered by any given consumer good with the appropriate price is complex enough. For the modern consumer looking to buy any given consumer durable – a car, or household equipment and furniture – with the lowest possible environmental footprint faces even greater complexity.

Washing machines now often come with eco-labelling for water and power use. However, the real issue from an environmental perspective might be where the machine is used (electricity has a vastly different environmental impact from one country to the next), as well as how often the machine is used and what sort of washing powder is used. For products containing wood – everything from paper to wooden garden furniture to household furniture – a non-profit organization called the Forest Stewardship Council (FSC) certifies wood, paper and other tree products that have come from sustainable forests.[2]

In practice, there are not enough such schemes, so the gathering of enough information to make an ecologically rational decision is impossible for even the most devoted environmentalist. *The Rough Guide to Ethical Shopping* suggests buying by brand, buying local or buying less as three approaches to ethical shopping – but if these lead to better choices from the perspective of the environmentally conscious shopper, a significant element of luck is likely to be involved. Buying brands known for an environmentally careful approach to manufacture means depending on information delivered by a party also keen to sell the product. On the other hand, even the manufacturer of products depending on a complicated supply chain may actually not have access to enough information to understand or report on a complete environmental footprint. If this sounds absurd, a brief look at some of the techniques used by manufacturers to assess the environmental impacts of their products will show why this can be the case.

Consumer durables make your head hurt – complexity and the consumer

An idea of the complexity inherent in assessing the environmental impact of products can be gained from considering a product's lifecycle. Lifecycle analysis is an approach to the measurement of the environmental impact of individual products or processes. The starting point is to establish the 'flow' of inputs. The 'flow' relevant to the car, for instance, would involve the following manufacture and usage stages:

Manufacture >> operation >> petroleum refining >> maintenance >>
fixed costs

Each of these stages is under the control of (and the responsibility of) different people or groups. Moreover, each of these stages can be broken down in far more detail. A lifecycle analysis just for a steel fuel tank in a car, for instance, would entail looking at tank manufacture, use and disposal. Tank manufacture involves carbon steel sheet, shield material, stamping and trimming, transport, galvanizing and coating, packing materials, paints, bearings, detergents for washing tanks, and lubricants and coolants. Disposal includes the inputs to shredding (electricity, transport, shredder tools) and steel recycling (inputs include limestone, refractories, electrodes, ferroalloys and electricity). Moreover, the precise impact of each of these smaller stages in a vehicle lifecycle assessment will depend on precisely what technology is used.

This gives some idea of the complexity of understanding the impact of car choice on the environment. The decision for the consumer can be simplified because by far the largest part of the environmental impact of the car is fuel used in the lifetime of the car. With cars, the problem with relying on consumer behaviour to reduce the impact of consumer durables on the environment is that each consumer is not responsible for one 'vehicle lifecycle' in which the car is made, used and then consigned to the equivalent of the knacker's yard, but several.

For other products, life is not so simple (not that it is that simple for the car purchaser). Something so seemingly basic as flooring finds a list of criteria relating to wood production (forestry standards), production of fibres (pesticides), production of foam rubber (harmful substance controls), flooring production (harmful substances, volatile organic compound (VOC) emissions, energy and resource consumption), usage (duration, cleaning requirements) and disposal (whether recyclable or not) (Bauman and Tillman, 2004, pp266–267). With flooring, the issues are spread across a broad range of elements in the supply chain. All the consumer can do is rely on 'eco-labelling', most of the time probably taking it at face value and with no real understanding of what it means.

If all of this sounds slightly mind-blowing, that is exactly the point. Every consumer durable we purchase is (in contrast to the household and transport equipment used by our ancestors) similarly complex. Whatever tool is used to understand the issue, exactly the same practical issues impede analysis.

What this means in practice is that the ordinary consumer must be dependent on those involved in the detail (the manufacturer) for enlightenment. For the consumer, the environmental impact of consumer durables is impossible to understand and therefore also impossible to control. The reality is consumer choice is extremely limited when it comes to environmental issues. Of course, in the long run, technology and human inventiveness might come to the rescue, providing a new generation of products that do less environmental damage in use – as in the case of the waterless washing machine or modular manufacture permitting a significant portion of any given consumer good to be recycled with ease.

What this survey of the waste landscape in electronic products demonstrates is what a huge amount of energy seems to be going into the waste problem. In fact, this could be framed as a materials conservation opportunity. As the old saying goes: 'Where there's muck, there's brass.'

Designing for de-manufacture

The general idea of 'sustainable consumption and production' denotes the 'fostering of lifestyles that are much less dependent on natural resources. This implies that goods and services need to be both produced more efficiently and consumed differently' (Wiedman et al, 2006). What that really means is that there is a need for a complete rethink regarding how things are made.

Xerox is famous for its remanufacturing system for used photocopiers. In the first iteration, based on remanufacturing entire machines, waste and carbon dioxide emissions were reduced by between 19 per cent and 27 per cent. In the second iteration, where remanufacture breaks old machines down into modular components for reuse, waste reduction over the lifecycle of a machine is between 38 per cent and 68 per cent, depending on the model. The question is why there is so much focus in regulatory discussion on how to dispose of waste, when finding ways of avoiding waste altogether would make far more sense.[3]

Obsolescence in some electronic products is designed in, and in others it is a consequence of technology developments. For the first category, designing out obsolescence and moving the business model to repair and maintain would make sense from an environmental perspective. For the second category, in which communications and information processing equipment are a case in point, modular designs making it easy for consumers to insert a new chip while keeping the basic machine otherwise unchanged could potentially reduce the flows of electronic waste.

Changing a couple of components rather than changing the entire piece of equipment could also be interesting from the perspective of the consumer trying to balance purchasing decisions within a constrained budget. Which brings us onto the financial credit crunch.

The financial credit crunch and consumer durable goods

So, how does economics interact with the consumption of consumer durable goods? We return to the four critical factors to be weighed up by a consumer in deciding whether or not to purchase a durable good: price, income, credit and the desire to purchase the durable good in the first place. (This final factor is a significant consideration, as we shall see.)

The right alignment of these various factors will lead to a consumer purchasing a consumer good. They do not all have to align perfectly. Consider Willy Loman, the eponymous salesman in Arthur Miller's *Death of a Salesman* (1949). Willy has a powerful desire to consume. He lacks the income necessary to make an outright purchase. The prices of the goods he wants to purchase are too high. But, critically, he has access to credit – and thus purchases a refrigerator, car and other consumer durables with instalment payments (completing the repayment programme only as these items reach the end of their working lives, of course). So, how does the financial credit crunch impact each of these factors? As always, the impacts are not consistent.

The financial credit crunch, income and consumer durable goods

The economics

As consumer durable goods have a relatively high price, the income a consumer earns is going to be important in determining what consumer durable goods they buy. The financial credit crunch has a variety of consequences for income, most of them negative.

The rise in unemployment during the financial credit crunch is one obvious negative driver of incomes. For someone unfortunate enough to lose their job, income all but ceases. No job means no income, which generally means no new washing machine. If the job loss is temporary, and employment is eventually regained, then consumption of durable goods will suffer little more than a hiatus. However, at least some of the unemployment caused by the financial credit crunch is going to be more enduring. There has been a rise in long-term unemployment (people out of work for more than six months) in the UK and elsewhere. One economy that has suffered more than most in terms of long-term unemployment is the US. Long-term unemployment in the US is now approaching continental European levels.

Economists know that the longer one is out of work, the less likely it is one is going to find a job – and long-term unemployment can easily morph into permanent unemployment. This is because the longer one is out of work, the more out-of-date one's job skills become. Employers are more inclined to think that there is something 'wrong' with someone who has been out of work for a long period of time. Finally, of course, unemployment is a demoralizing experience, and those out of work for long periods find it difficult to motivate themselves to find the work they are capable of doing.

Furthermore, the financial credit crunch has also taken its toll on public finances. Indeed, in most Organisation for Economic Co-operation and Development (OECD) countries the state of public finances is something that should carry the warning 'this government budget contains scenes some viewers may find distressing'. Ultimately, there will be more taxation to meet funding requirements (maybe even to pay back some of the debt). Taxation (whether in the form of an income tax or a sales tax) will reduce the ability of the individual to spend money. This constraint applies to all goods and services, of course, but to the extent that income is a more important driver of the consumption of 'big-ticket' (high-priced) consumer durable goods, this is particularly relevant.

So what does this mean? It means for a section of the population (the lowest income groups, in effect) the ability to consume durable goods will decline significantly – possibly entirely. For a majority of the population, experiencing a more constrained income (or at the very least, a more constrained pace of growth in that income), the ability to consume durable goods will be diminished. Only the highest income groups are likely to continue to consume durable goods at the same pace as prior to the financial credit crunch – because consumer durable goods are a relatively low proportion of their income. Even here, however, consumer psychology may work against their purchasing at the frenetic pace of the past two decades.

The environment

What might this mean from the perspective of environmental impacts? We suggested that the environmental impact of consumer durables could be reduced in three ways: by designing them so that fewer resources (such as water and energy) are required to use them; by using them as their names suggest they should be used, for long periods of time; or by recycling them or their component parts. The income effects of the credit crunch suggest that consumers would only purchase environmentally efficient consumer durables if they were cheaper than the usual equivalents; or that they might well (by necessity) be forced to use durables for longer, or recycle. Furthermore, should recycling prove costly (in time, money or effort), disposal would be the more likely course of action for items that could not be sold on.

The financial credit crunch, goods prices and consumer durable goods

The economics

The impact of the financial credit crunch on the price of consumer goods is a very subtle impact, but there are some potential transmission mechanisms. However, the reaction of durable goods prices to the financial credit crunch has not been uniform. In the UK, the price of durable goods has tended to rise (relative to other goods and services). In the euro area, durable goods prices have had deflation in absolute terms and fallen in relative terms. In the US, durable goods prices have tended to rise in line with the average inflation rate for the entire economy.

So what are the interrelationships between the financial credit crunch and the price of consumer goods? Moreover, why have the price experiences been so different?

There is a circular argument in the pricing consideration, of course. If demand for anything is weak, then the price will tend to decline. If the demand is robust, then the price will tend to rise. The price–demand relationship is symbiotic. This may explain why durable goods prices have held up in the UK; other factors have facilitated a stronger level of demand (particularly the credit cycle), which means that pricing has been relatively firm. However, it is worth considering that this price impact is only likely to last as long as existing stocks of supply last. If demand for cars is weak in the US, for example, the price of cars will tend to come down (at least in relative terms). However, car producers are not going to carry on producing cars in the face of weak demand – even American car producers have a better business model than that. Once the inventory of cars has been used up, the reduced demand will be met by reduced supply, and prices will be under less negative pressure. (Inventory is what the BBC's *Blue Peter* programme would describe as 'here's one I made earlier'. It constitutes goods that have already been manufactured, and are simply sitting around waiting for a buyer to materialize.)

However, it is important to remember that the price of consumer durable goods is only one factor determining demand patterns. Before the First World War, the relative price of consumer durable goods rose significantly in the US, but demand also increased. Other factors were at work. However, in the 1920s, consumer durable goods fell in price relative to other goods (i.e. the inflation rate for consumer durable goods was less than the inflation rate for other goods) though they did not reach the price levels of the pre-war era. Demand for consumer durable goods also rose. Indeed, economists often refer to the 1920s as the 'Consumer Durable Revolution'. As descriptions go, it is not perhaps as evocative as the 'Roaring Twenties', but it conveys the same message.[4]

Three other factors need to be considered when looking at the financial credit crunch and consumer durable goods' prices: labour costs, capital costs and international prices.

Labour costs matter because (generally speaking) labour costs form an important part of a consumer durable good's overall cost base. If labour costs are rising, then the labour that goes into durable goods will push up the price of those goods. Moreover, because the labour costs are generally higher-skilled labour, it is the wages of a specific area of the workforce that we are concerned with. The impact of the financial credit crunch to date has been to raise unemployment and suppress wage growth. If this continues to be the case, then the impact is likely to be a moderation of consumer durable goods' prices.

Capital costs are important because consumer durable goods tend to be more capital intensive than non-durable goods (and are obviously more likely to be capital intensive than is the service sector). Think of the amount of machinery it takes to produce a car, for example. That machinery must be funded somehow, and it is normally financed through credit. If credit becomes more expensive (and it has become more expensive in the wake of the financial credit crunch) then the cost of installing machinery rises, and thus the cost of the end product – the consumer durable good – will rise. As politicians increasingly insist on the regulation and taxation of credit, so the cost of consumer durable goods is likely to rise.

Finally, international pricing is also important to consumer durable goods. Consumer durables are generally highly traded goods. More than 90 per cent of the world's washing machines are manufactured in China, for example. If the financial credit crunch leads to restrictions on international trade or imposes tariffs or affects international prices in some other way, then this has a bearing on durable good demand. Americans who want to buy a tyre for their SUV are likely to face higher prices than would otherwise be the case, because the American administration has chosen to tax consumers of foreign (specifically Chinese) manufactured tyres through the medium of a tariff. The motives behind the decision to impose a tariff are complex, of course. However, most economists would at least ascribe some of this rush to protectionism as an unfortunate consequence of the financial credit crunch.

So, does this mean that the impact of the financial credit crunch is to increase or to reduce the price of consumer durable goods? The case is, confusingly, not proven. The balance of probabilities is that the price of consumer durable goods is unlikely to rise exceptionally. Specific consumer durables may increase in price (particularly if subjected to protectionism), but prices are not *generally* likely to be higher.

The environment

Following on from this, where are price effects likely to be most relevant in the area of the containment of the environmental footprint of consumer durables? Redesigning goods so that they need fewer resources to make or so that they last longer can put up costs but does not have to. Indeed, there are consumer durables manufacturers that sit in an environmental–economics 'sweet spot', able to deliver more environmentally-efficient goods to consumers more cheaply, as well as making a decent profit margin. (As an example, any firm making significant improvements in energy efficiency, reducing its carbon footprint and energy bill at the same time, would be in this position.) Firms with the wherewithal to manufacture goods on a modular basis may be able to offer consumers the choice of changing parts so they can easily exchange and plug in rather than repurchasing the entire thing. Failing this, if costs cannot be brought down, a fall-off in the turnover of consumer durables is the most likely route by which environmental impacts might be reduced. As discussed above, unless accompanied by a cultural change, this will be temporary, assuming the grip of the credit crunch does not take a very long time to loosen up.

The financial credit crunch, financial credit and consumer durable goods

The economics

Self-evidently, credit is where the financial credit crunch leaves its deepest scars on the global economy. There is less credit available. What credit is available is more expensive. And people who had access to credit in the past are going to find that their access to credit is curtailed in these more austere times. None of these are temporary phenomena (although, admittedly, they might be temporarily exaggerated by more extreme caution on the part of banks and other suppliers of credit).

Credit costs are rising. Banks are charging higher interest rates because they see higher risks, and they want to be compensated for those risks. This applies across all forms of consumer credit – from mortgages through to credit cards. Even as US official interest rates fell to zero in the wake of the financial credit crunch, the average interest rate levied on American credit cards rose. Clearly, a higher credit cost is a disincentive to borrow.[5] Credit in the next decade is likely to be more expensive than it has been in the past decade. The impact that this has on purchases of consumer durable goods is likely to be noticeable.

Limiting credit in an absolute sense is also a critical driver of demand for consumer durable goods. In the 1920s, 64 per cent of new cars purchased in the US were acquired on instalment terms, and 40 per cent of American households purchased a car during that decade. However, the UK was far slower to acquire cars than the US. This was partly a function of price (cars were more expensive in price terms in the UK), but also a function of credit availability.

British finance houses were, generally speaking, more cautious than their American counterparts about *whom* they would offer credit to. Many British professions lacked a regular income (a private sector doctor paid in consultancy fees does not have a regular salary, of course). This was seen as an obstacle to gaining credit in the UK, but not in the US. Credit was less readily available in the UK, and thus the purchase of consumer durable goods in the form of cars was less readily available.[6]

The US, meanwhile, was indulging in an orgy of borrowing in the 1920s. In 1900, non-mortgage consumer debt amounted to less than 4.5 per cent of disposable income. By 1930, it was more than 9 per cent. Credit became more widely available, at a lower cost (and generally over a longer period). Offered credit, Americans took it. Only with the onset of the financial crash of 1929, which was a financial credit crunch of course, did this availability of credit dry up and frugality start to re-emerge.

As economies evolved from the frugal era of the early 1930s to the age of self-certifying mortgages, credit became more widely available. Consumers who would not previously have been considered eligible for credit, suddenly found themselves inundated with offers from finance houses. Credit was made available to people with lower incomes, in particular. The 1990s saw a particularly aggressive expansion of consumer credit, especially in the Anglo-Saxon economies. This facilitated an increase in the purchase of consumer durable goods.

The sort of credit that was available also shifted. In the 1980s, the credit offered to lower-income groups in society was generally confined to instalment credit – in a form largely unchanged over the preceding 70 years. Regular, predetermined and fixed payments would be made over a specified period if one wished to buy something on credit. That gave certainty to the consumer as to their monthly budget. In the 1990s, this changed, and instalment credit began a steady decline. Instead of instalment credit, credit cards became the preferred source of credit. This shifts the debt from fixed interest rate to floating interest rate, and also removes the necessity of paying off debt over a rigid time horizon. Buying a washing machine with six monthly instalments means the debt is cleared within six months. Buying a washing machine using a credit card means that the washing machine is paid off over a far longer time frame (20 months, or even more, depending on the terms of the credit card).

So far, the credit implications of the credit crunch seem to reduce the demand for consumer durable goods. For many, particularly lower-income groups, this is undoubtedly the case. Acquiring or upgrading consumer durable goods will be more difficult in the future than in the heady consumerism of the past two decades. Lower-income groups are, in effect, reliving the 1930s. The US credit boom of the 1920s saw more than 15 per cent of households buy a car *using credit* in the US in 1929. For a single year, this is a truly remarkable figure (even more remarkable, nearly a quarter of all households in the US bought a car that year). Moreover, more than 10 per cent of households a year had been using credit to buy a car since 1923. (In actual fact, in 1927, only 9.8 per cent of Americans used credit to buy a car. This temporary moment of restraint does not really disrupt the central argument, however, and so we round up for effect.) Clearly, the recent credit-fuelled hedonism of the US consumer may owe something to genetics. After 1929, the availability of credit slumped. By 1933, 4.2 per cent of households were using credit to buy a car.

No credit, no car. It was that simple. And history seems to be repeating for the great-grandchildren of the 1920s generation.

So much for the fate of lower-income groups in a credit crunch. The situation is not so clear for higher-income groups.

Higher-income groups are affected by the more credit-constrained environment. The rise in the cost of credit and the reassessment of risk premium is universal. Although higher-income groups will be distinguished by more favourable credit terms (indeed, the gap between their credit terms and those of their poorer contemporaries is likely to widen), in an absolute sense the cost of credit is likely to increase for higher-income groups. However, higher-income groups are less likely to need credit to purchase consumer durables. A £300 washing machine represents 72 per cent of a month's gross income for a household in the bottom 20 per cent of income distribution; it represents only 4.9 per cent of a month's income for a household in the top 20 per cent of income distribution (2008/2009 gross income levels). The situation today is not so far removed from the comparable affordability of the mahogany dining room table of Georgian times.

The environment

The main message of the economics of the credit crunch with regard to credit appears to be a fall-off in volume demand and, therefore, a proportionate fall-off in the environmental impacts of the same. Contrasting that, there is a potential negative from constrained government credit. Solar energy been installed in social housing in the UK, with support from the local authority. The resident of the flat pays an electricity bill that is a bundle of electricity used and interest on the loan. In aggregate, this is structured to be slightly lower than the usual electricity bill. This is effectively government-sponsored credit with a positive environmental outcome. However, if the cost of credit (or the availability of credit) for the local government is constrained – and it is – then the prospects for such a scheme look bleak. This then hits the lower-income consumer twice – they are denied the opportunity to make an 'environmental upgrade' by the private sector and the government sector no longer has the resources available to be able to substitute for the absent banking system.

The financial credit crunch, the desire to buy and consumer durable goods

The economics

A financial credit crunch will do odd things to consumers. One thing that has been noticed in the current economic climate is the resilience of the sale of home furnishings and major electrical appliances. Indeed, furniture pricing has also managed to hold up reasonably well.

The desire to purchase durable goods is complex. There is an element of social aspiration – but this affects all forms of goods. Indeed, one could argue that the social trends of the 1990s onward have been towards expressing social aspiration through consumer non-durable goods (if we count clothing as such) – for the younger generation, the latest designer brands are the symbol of social status – be they clothing or consumer electronics. One of the authors occasionally speaks to schoolchildren about careers in the City of London and there are two questions one is certain to be asked, no matter what the audience age. 'How much do you earn, sir?' is the first. 'What sort of phone do you have, sir?' is the second. The answer of 'a standard work-issue Blackberry' does not tend to engender a great deal of respect from the audience.

More importantly, to a culture that is deeply immersed in the necessity of consuming, the desire to consume can be generated by the necessity of downsizing. This is akin to the consumer behaviour we discovered in our opening chapter on food: consumers reject the expensive meal out, but instead opt for an expensive ready meal and a bottle of wine at home (in the interests of economy). This trait applies to consumer durables as well, and it is a trait that is accelerated by the financial credit crunch.

One of the key characteristics of the credit crunch is that people are moving house less frequently. Moving house is generally expensive. The credit crunch may make it more expensive, depending on the mortgage structure. To move house in the UK will generally necessitate the negotiation of a new mortgage. If a borrower has locked into a mortgage in the pre-financial credit crunch era, the terms of that mortgage are almost certainly better than what the same borrower could achieve today; banks underpriced risk in the years prior to the financial credit crunch. Today, they are, belatedly, recognizing the importance of appropriately pricing risk.

What this means is that moving house is now more expensive than has been the case hitherto. Add to this the fact that consumers may have seen the value of their current home fall, and the volume of housing market transactions will fall – as indeed is the case.

What has this to do with the demand for consumer goods? It is simple. If a home owner feels unable to move house (and move a rung or two higher on the property ladder in consequence), then they will seek to 'trade down' and seek consumer gratification in a less expensive manner. If the home owner cannot afford to move house, they might choose to install a new kitchen, build a loft extension, add on a conservatory or update the three-piece suite.

This desire to upgrade one's property as an alternative to moving house is, clearly, focused on demand for consumer durable goods. By constraining (or making more expensive) mortgage credit, the consumer is encouraged to increase their purchase of consumer durable goods. This trend has been particularly acute in the UK, but has also supported consumer demand elsewhere. Clearly, other issues interplay with this demand for consumer goods. Again, income is important here. Whether the purchase of a new kitchen (or conservatory, or whatever) is viable depends on access to non-mortgage credit, pools of savings and overall income levels. However, if the desire to consume is set in motion, it certainly pushes the balance of probability in favour of increased consumption of consumer durable goods.

The environment

The evidence suggests that, as consumers have become increasingly aware of their environmental impact, the desire to incorporate the ecological impact of products alongside other benefits and services delivered by the product is growing. If the financial credit crunch continues long enough to produce deflation across a broad range of goods, people will tend to defer 'trading up' to new technology, and this (although negative in many other ways) would offset some of the environmental impacts of Moore's Law.

The financial credit crunch and consumer durable goods

The financial credit crunch has had a complex impact on demand for consumer goods. Some focus may be achieved when viewing the position through the lens of income levels, however. Generally speaking, lower-income groups are less and less likely to purchase consumer durable goods. Income constraints and credit constraints will outweigh any other factors affecting the decision to purchase.

For middle and higher-income groups, demand for consumer durable goods is less likely to be constrained. The income constraint is still there, but it is a less binding constraint for many. Credit is more expensive, but that is not an overwhelming constraint (and credit is likely to be available – whereas for lower-income groups the financial credit constraint may mean that credit is not available at any price). Most significantly, middle and higher-income groups may choose to consume durable goods as a cheaper substitute to moving house.

This means that society splits. Lower-income groups will, in the absence of policy assistance, continue to use existing consumer durable goods. The most environmentally positive option in this set of circumstances is that they purchase the 'old' consumer durable goods that their better-off neighbours discard. This at least reduces the environmental costs associated with the manufacture of consumer goods. However, the operation of these consumer durable goods is likely to come at a higher environmental price than modern technology can deliver. In other words, over time, the financial credit crunch worsens the environmental credit crunch for lower-income groups, and Earth Overshoot Day moves nearer rather than further away.

The higher one moves up the income chain, the more likely it is that consumer durable goods will be replaced frequently. New cars may be a status symbol, but they also increase energy efficiency. Modern kitchens will tend to be more energy efficient than a kitchen installed ten years ago. Indeed, there is now a change in the European Union's (EU) 'energy-efficient' rating for domestic appliances, with A+, A++ and A+++ added to the benchmark. So many appliances achieve the 'A' grade for energy efficiency (according to the standards determined 18 years ago) that it was decided that additions to the scale were needed, coming into force from July 2011. Around 90 per cent of refrigerators, washing machines and dishwashers now achieve an A grade, which makes distinguishing one from another impossible (and, of course, reduces the incentive for manufacturers to strive for greater energy efficiency).

The cost of this, in environmental terms, is that old consumer goods may be scrapped well before their economically useful life is over. The environmental efficiency versus environmental capital cost balance is a difficult one to determine. It is also something that economics is unlikely to contribute too much to in the current climate, as the aftershocks of the financial credit crunch distort the smooth operation of the market.

In a nutshell, the ability to consume durable goods is likely to be diminished by the financial credit crunch. Only the highest income groups are likely to continue to consume durable goods at the same pace as they did before the financial credit crunch. However, if frugality became more fashionable in the aftermath of the financial credit crunch, then recycling and resource stewardship could become the new collective good, even for the relatively wealthy.

As things stand, most of the green consumer 'models' look for products containing ecologically sourced materials in combination with a focus on recycling. This essentially maintains consumerism as the dominant model, and this goes hand in hand with the traditionally calculated gross domestic product (GDP). Will we continue to depend, for economic growth, on the throwaway model of consumerism as markets move out of the financial crunch? The chances are that, in the short run, governments will welcome an upturn in spending whatever the source and regardless of how environmentally unfriendly. Further out, the shock of the crunch may have changed thinking about spending and durables. Even further out, the shock may have helped propel forwards a cultural change we see as currently embryonic; 'greening' consumer durables will produce consumer durables people desire, bringing many opportunities for forward-looking consumer firms that can deliver them at the right price.

Notes

1 The majority of households own most of these things, with exceptions – for instance, only a third of lower-income householders own a computer.
2 The ten principles used to judge this include uncontested long-term land tenure, equitable use and sharing of benefits derived from the forest and maintenance of High Conservation Value Forests.
3 For more on the Xerox story, see Kerr and Ryan (2001).
4 The development of the consumer in the US in the inter-war period is covered in Olney (1991, p56 looks at the relative price shifts detailed here). Bowden and Turner (1993) also cover the same period with US/UK comparisons.
5 There is an exception at the extreme. The most desperate in a society will tend to borrow 'at any price'. In economic terms, they have '(credit) price inelastic demand'. In other words, if you are desperate, you will go to a loan shark and borrow at interest rates you know you cannot afford.
6 This section is drawn largely from the research of Bowden and Turner (1993).

References

Bauman H. and Tillman A. (2004) *The Hitchhiker's Guide to LCA: An Orientation in Life Cycle Assessment Methodology and Application*, Studentlitteratur AB, Lund, Sweden

Bowden, S. and Turner, P. (1993) 'The demand for consumer durables in the United Kingdom in the interwar period', *The Journal of Economic History*, vol 53, pp244–258

Clark, D. (2004) *The Rough Guide to Ethical Shopping*, Rough Guides Ltd, London

Kerr, W. and Ryan C. (2001) 'Eco-efficiency gains from remanufacturing. A case study of photocopier remanufacturing at Fuji Xerox Australia', *Journal of Cleaner Production*, vol 9, pp75–81

Larsson, M. (2009) *Global Energy Transformation*, Macmillan, London

Light, A. (2007) *Mrs Woolf and the Servants*, Penguin Books, London

Olney, M. (1991) *Buy Now, Pay Later*, University of North Carolina Press, Chapel Hill, NC
Vickery, A. (1998) *The Gentleman's Daughter*, Yale University Press, New Haven, CT
Wiedmann, T., Minx, J., Barrett, J., Vanner, R. and Ekins, P. (2006) 'Sustainable consumption and production: Development of an evidence base', Stockholm Environment Institute and Policy Studies Institute, Stockholm and London

CHAPTER 7

Fast-moving Consumer Goods: 'Waste Not Want Not'

Once consumer aspirations are awakened, a lack of economic means blocking access to the 'heavier' objects of desire will redirect that interest toward the 'lighter' objects. (De Vries, 2008)

Fast-moving consumer goods (FMCGs) are traditionally defined as relatively frequently purchased goods. The main difference with the durable goods products described in the previous chapter is that fast goods are consumed, break, wear out or become obsolete when used as intended. Economists, equity analysts and environmentalists probably disagree on what should constitute an FMCG. Having the broader definition, economists are likely to be the ones with the right answer in describing how FMCGs are defined in the here and now. (Economists are also likely to be right because economists normally are right. It is just that the world is occasionally out of sync with economic forecasts.)

As the chapter on consumer durables also suggests, an expanded modern-day definition of the FMCG, including the emergence of the fast-moving 'durable' consumer good (perhaps creating a potential new grouping, FMDCG), may lie at the root of some of the environmental issues we address in this book. To avoid confusion, we use the term 'fast goods' to refer to a broad category of FMCGs.

On this broad definition, a fast good is basically that which is not a consumer durable (but also not food or energy). When economists look at economic data, they tend to distinguish between food, energy, consumer durables and other consumer goods. This last category is the fast good grouping. Equity analysts tend to be somewhat narrower in their view (that equity analysts have only a narrow vision will occasion no surprise whatsoever to economists). Equity analysts' analytical scope currently tends to encompass only the companies that manufacture food and household products. Leaving the niceties of the economist versus analyst debate to one side, the environmentalist would argue that the slower the merry-go-round of consumption, the better it is likely to be for the environment (in the sense that the 'slow' consumer basket is, other things being equal, likely to be less resource-intensive).

For the environmentalist, it is therefore useful to take the current *economic* definition of what a fast good is, because this helps to throw the durable/fast good problem into relief. Taking the economic concept, the fast goods category encompasses clothes (today – as we saw in the previous chapter, this has not always been the case). It includes some electronics and some forms of delivering entertainment. (The entertainment itself is generally a service, not a good.) Thus an MP3 player may be a fast good. The music downloaded to the MP3 player is a form of cultural service and would not be classed as a fast good. Children's toys are part of the fast good category, as (at the speediest end of the fast good spectrum) are certain household and personal products – cleaning products, toiletries and so forth.

The key point is that the expansion of what constitutes a fast good has happened in parallel with the expansion of the average environmental impact implicit in the typical overall consumer basket. Both have happened as a result of development of new products and concepts (for example in entertainment), together with changes to manufacturing processes and marketing (the latter being often responsible for shaping what the consumer wants along with the environmental profile of his or her shopping basket), together with changes to consumer behaviour (driven by the desire for a better quality of life among other things). The implication is that a change in all of these (not necessarily a reversal) may be required to reduce the environmental impact of the consumer.

The typical environmentalist would be in favour of a reduction of the number of fast-*moving* goods in the fast goods basket or, at least, a reduction in the speed of movement thereof, for the good of the environment. The question we address in this chapter is whether the credit crunch could put the brakes on some fast goods and send them (or send them *back*) into the durable goods basket. As we will see, what this chapter suggests is that the effects of the financial credit crunch on fast goods (all things being equal) are unlikely to slow down the 'fast good' whirligig so as to meaningfully impact the environmental credit crunch. The only hope is that other things prove not to be equal and other catalysts such as cultural change or innovation come into play. Perhaps rather perversely, the driver of many of the more durable fast good purchases – fashion – holds the most potential for such a change in purchasing patterns.

The economics of fast good demand

The economics of fast good demand are rooted in a mix of need, fashion and envy (or if we wish to avoid the pejorative language of the deadly sins, 'aspiration'). The way in which the financial credit crunch changes these motives for demand will shape the way in which the credit crunch interacts with the fast good sector. Fast goods will not react in the same way as other sectors of the economy. Unlike food, for example, many fast good purchases are not high-frequency (weekly) purchases – with the exception of personal care and household cleaning products, fast goods are distinct, one-off purchases rather than part of the weekly shopping basket. Even personal care products are often bought at fairly infrequent intervals. Unlike consumer durables, something that falls into the fast good category is unlikely to form a significant proportion of the household budget (though this does not preclude credit being used for the purchase of fast goods).

Perhaps against expectations, fast goods that cannot be regarded as daily necessities (and therefore could theoretically speaking be forgone more easily to save money) tend to be things that people cling to during a financial credit crunch. That peculiarly British expression 'all fur coat and no knickers' captures this. Because, in a relatively materialistic world, people regard possessions as part of the definition of their status and their standard of living, they are reluctant to give up on the outward display that fast goods represent. People are very reluctant to feel that their standard of living is falling, and even more reluctant to have their neighbours perceive that their standard of living is falling.

Fast goods represent this, in a way that service sector spending tends not to. It is a very common trait in an economic downturn that consumer spending will fall more rapidly than retail sales. Consumer spending is all spending, and includes spending on services. Retail sales represent the purchase of physical goods.

The drivers of the fast good environmental impact

Economics suggests that the components of demand for fast goods are driven by something other than price, and may be driven by something other than necessity. Thus, the impact of fast goods on the environmental credit crunch is even further removed from the conventional pricing forces of macroeconomics.

In essence, there are four dimensions to the way fast goods affect the environmental credit crunch: the resource intensity of the consumer basket of fast goods;[1] the environmental impact of certain potentially toxic materials used in fast goods and their packaging; the quantity of fast good products consumed by the typical household each year; and how much of the fast goods ends up as waste at the end of the product's life.

These can be broadly thought of as two pairs. The resource intensity and environmental toxicity are about what goes into a product; the volume and the waste issues are about the consumption of a product – and, of course, the issues of volume and waste are often highly correlated in a 'disposable' society.

Sources of fast good demand

The interplay of economic shifts in need, fashion and envy with the four environmental issues surrounding fast goods is the key to how fast goods will react to the twin credit crunches. Price, income and an unwillingness to return to 'old' ways of doing things all have the same role in this interaction. The former are features that are affected by the financial credit crunch, while the latter is essentially a social issue.

In combination, the economic and environmental forces suggest that reducing the environmental footprint of fast goods could be achieved in the following four ways:

1 By improving the environmental efficiency of the procurement and production processes.
2 By redesigning the product in one or more of the following ways: to reduce the amount of product needed by the consumer (for example, by making it last longer); to remove or reduce the environment-altering substances in fast goods; to enable fast goods (or their packaging) to be reused or recycled rather than thrown away.
3 By influencing consumers to use fast goods in such a way that their impact on the environment is reduced – for instance, through labels on the packaging.
4 By influencing the design of those consumer durable goods that use fast goods (for example, washing machines and detergent) so as to change consumer behaviour (for example, using less detergent).

Fast goods in the 21st century are complex beasts. For a fast good supplier, they can be seen as the bundle of manufacturing process, intellectual property and brand, shaped by a number of forces – social trends, advances in technology and microeconomics. For a consumer it is all about need, fashion and envy.

The interaction of fast goods with the financial credit crunch and with the environmental credit crunch is likely to be unpredictable and could run either way. This is a pity – the fast good category, more than any other perhaps, defines what the modern consumer society is about.

The definition of need (and whether it might expand or contract with the financial credit crunch) becomes fundamentally important to understanding where fast good demand is likely to go, and so we explore this a bit more in the next few paragraphs, before going on to enter the world of fashion.

The financial credit crunch and the need to use or own fast goods

'Need' is an elastic concept. In its most primitive form, humanity needs water, food and some form of shelter (in that order of priority). However, as economic society has developed, so has the perception of what it is 'needed' (in material terms) by members of society. In the 1970s, a colour television was seen as a luxury item and many families did not own one. By the 21st century, a colour television was seen as such an indispensible part of the minimum acceptable living standard that using government-funded social assistance to purchase one was considered acceptable.

The 'need' concept for fast goods therefore depends not just on what is required to sustain life, but what society perceives to be the minimum acceptable standard of living. This clearly changes over time and between economies. One complicating factor can be the rise of income inequality that was a hallmark of the financial credit expansion in many countries. If income inequality increases, then the average standard of living becomes impossibly beyond the means of the lowest-income groups in a society. Nevertheless, the need for fast goods (dictated by marketing and other factors) will lead society to conclude that the minimum amount of possessions needed is some proportion of the 'average' income in the society. It is perfectly reasonable to suppose that, in particularly unequal societies, more than half the population is below the perceived 'need' level for fast goods.

The economics

The 'need' to have certain goods breaks down further. On the one hand, there are the 'super-fast' fast goods. These are things like household cleaners, toiletries and the like. On the other hand, there are the 'slower' fast goods – consumer electronics, children's toys and similar products, which are not really consumer durables but equally are bought with less frequency than the super-fast fast good sector.

The financial credit crunch has generally had the effect of shifting demand patterns somewhat when it comes to 'needed' fast goods. For the super-fast category, there will be an attempt to economize on the amount spent, though not necessarily the volume consumed. For the slower fast goods, there is likely to be an attempt to increase the durability of the 'needed' product. This might then push some fast goods into the durable goods category or, at the very least, push some fast goods in that general direction. It is extremely unlikely that the financial credit crunch will change the perception of *what* is needed.

The financial credit crunch is constraining the total spending power of the consumer. This comes in three ways: first by reducing the pace of economic growth (which, all things being equal, will slow the growth of personal incomes); second, by raising taxes in the medium term (to finance higher debt burdens); third, by reducing the amount of credit that a consumer has access to. This combination, as we have already seen, shifts demand for high-frequency purchases such as food. It will also constrain credit-dependent larger purchases such as durable goods. However, because consumer credit had become so important in the run up to the financial credit crunch, constraints on credit will also have an impact on the fast good sector of the economy. If a household is seeking to reduce its credit card debt, for instance, the ensuing economies will inevitably curtail consumption of a wide variety of goods.

So what does this mean for the 'need' aspect of demand for fast goods? Because consumers cry out with the petulance of a toddler, 'but I *need* it', demand for those goods that fall into the fast good 'need' category is unlikely to fall just because of a little thing like a financial credit crunch. Life would be insupportable without such items (in the mind of the consumer). To reduce the 'need' component would be to accept a decline in one's standard of living (or, at least, a decline using the metric that society adopts to measure the standard of living, i.e. material possessions).

There is something of a ratchet effect here. To quote Joni Mitchell, 'you don't know what you've got till it's gone' (the song was *Big Yellow Taxi*, 1970). Consumers who have not known the joy of owning a flat-screen television (or, indeed, the benefits of using deodorant) may not regard ownership or use of such a fast good as an absolute need. A consumer who has owned a flat screen television for any period of time will refuse to contemplate existence without one. To surrender the necessity of owning or using such an object would be to accept a decline in one's standard of living.

With the super-fast fast goods, the consumer may try and reduce waste. Careful measurement of the washing powder, a more conservative application of moisturizer and so on, could be part of an attempt to consume only what is needed. However, such waste control is unlikely to endure, not least because the cost is not particularly visible (does the typical consumer really throw up their hands in horror if an extra 10ml of fabric conditioner escapes the bottle?).

For the more enduring fast goods, the issue is not immediately a question of changing the quality of goods purchased (though we will come to that). The very name 'fast' goods implies a certain impermanence to such goods – and this presents a dilemma when these objects reach the end of their natural life. The necessity of ownership requires that the consumer has these goods in their possession, but the financial credit crunch argues that the resources with which to replace them are less readily available.

In general, the financial credit crunch leads to an attempt to prolong the life of the more durable fast goods, if possible. Repair rather than replace becomes the guiding principle for those fast good items that are purchased principally out of need. Thus, the financial credit crunch has been a driver of economic expansion for some industries. Spending on shoe repairs in 2009 *rose* 7.8 per cent in real terms in the UK, at a time when spending on the purchase of new footwear *fell* by 6.4 per cent. Shoe demand can be thought of as need-based (there is a fashion element too, to be sure).

Generally speaking, one repairs what one has, in the interests of (household) economy. The consumer is betting that the repaired object will last long enough that the cost of the repair will be less than the (proportionate) cost of replacing the faulty fast good. Alternatively, if the consumer is budget-constrained, he or she will be hoping that by the time the repaired object fails again, their income or credit position will be better and they will be able to afford a replacement more readily.

However, the durability question reverses when the consumer has to replace (because repair is not practical), or when a new fast good is added to the household stock of objects. In a financial credit constrained environment, the likely reaction is a trading down in terms of quality. If the television breaks beyond repair, but a television is a 'need' fast good product, then the consumer will tend to turn to the cheapest alternative available. This suggests that the future durability of the fast good portfolio will deteriorate.

The environment

We have already identified the four things that drive the environmental impact of the typical fast good item as the resource intensity of the product; the environmental toxicity of the constituents that make up the product; the volumes actually used; and what happens to the remains of the product or the product itself once it is no longer used at a time when 'waste not want not' has lost much of its meaning. The question is how these four factors relate to changes in 'needs' based demand for fast goods?

Fast goods, need, resource intensity and toxicity

For the fastest-moving fast goods, the reaction of the consumer to the crunch in the context of food provides a useful guide. The key characteristic of a high-frequency purchase is that it is, well, very frequently purchased. As such, the sensitivities to price come through. At the same time, one does not abandon the purchase of toothpaste (or food), no matter how frequent the purchase, just because of the financial credit crunch. To someone who has never used toothpaste, the absence may not be a problem. Once one has started on the route of dental hygiene, however, there is no turning back ('you don't know what you've got till it's gone' applies to toothpaste and, come to that, teeth, as well as televisions). Volume, in short, is not an issue.

What one does is trade down in terms of the quality (or perceived quality) of the purchase. Thus, we see an attempt to economize in terms of the amount spent on a product, while the product itself is still purchased. In environmental terms, this will matter if the lower-cost products are more environmentally damaging than the more expensive products (if the cheaper product is more resource intensive). The problem is that it is not clear in advance if trading down in quality results in a positive or negative environmental change (in terms of either resources used, or the toxic nature of those resources).

Box 7.1 Will trading down help or harm the environment?

The environmental consequences of trading down to lower (economic) cost fast-moving goods is impossible to predict because environmental considerations have not historically driven the development of the fastest-moving consumer goods – unless by accident. Nowhere is this better illustrated than in the history of domestic laundry. In Shakespeare's *The Merry Wives of Windsor*, Falstaff is put into a buck-basket and taken to Datchet-mead to be emptied into a muddy ditch. 'Bucking' referred to the process in which items to be cleaned were soaked in lye or, indeed, chamber lye. (Chamber lye is urine. It is worth considering that urine could be considered a typical 'fast good' in its day. Historically, far from being considered waste, it was a useful household product and kept in small tubs in the back yard for use in the domestic laundry.) Sometimes this process was followed by washing with soap, but not always.

Before modern detergents and washing machines, laundry was a hard and dirty business involving one or more of four processes: pounding, steeping in the harsh smelly substances, boiling or rubbing with soap. This would be followed by a complex labour-intensive process of bleaching, starching, mangling and ironing.

Looking back with rose-tinted spectacles, it is possible to see old approaches to washing as 'green' on the basis that the chemicals used for washing were both recycled and biodegradable. However, were we to contemplate trading down to the same cleaning materials today, the water pollution from the lye-based laundry of any urban city would be considerable. There would be significant problems with water algae – not to mention carbon emissions from burning ferns for their 'superior' ash. These were not valid concerns in the smaller, pre-detergent economy.

Unless environmental impacts are fully integrated with technological improvements designed to improve the quality of the relevant fast good items, social needs (such as the need for cleaner, less labour-intensive laundry) will dominate the development of fast goods. This leaves the environmental impact open to random chance as one trades up or trades down the quality of super-fast fast goods.

The trading down element is therefore something that might or might not influence the environmental resource intensity of the fast good sector. For the fastest-moving fast goods the point is moot. However, as we decelerate along the speed arc of fast goods (to children's toys or to basic consumer electronics, say), trading down is more and more likely to err on the side of environmental harm. This reflects the fact that trading down in quality will generally imply a shorter-lived product. The product itself may have the same environmental resource input as its better-quality peer but, because it is shorter lived, the volume purchased over any given period will rise. Hence, we move to the relationship between needed fast goods, and waste and volume.

Fast goods, need, volume and waste

'Make do and mend' was one of the many propaganda slogans of the British government during the Second World War. The thinking behind this was, of course, that by adopting a philosophy of 'repair rather than replace', fewer resources would be used by the civilian population (allowing more to be used in support of the war effort). Using fewer resources through consuming less will mean wasting less – which will, of course, lessen the environmental impact of consumption.

However, make do and mend has become a lot harder than it used to be because the fast good items of today are often complex, in order to make life easier for the user. With complex fast goods one can use one product for several applications. Such complexity is apparent in cleaning and washing (using super-fast fast goods) as well as in electronics and other 'slower' fast goods.

Consider the materials traditionally used to remove stains from textiles in the domestic household – *depending on the type of stain*, the following were applied as appropriate: boiling water, cold water, dry starch, salt, lemon juice, half of a lemon (as a rubbing agent), turpentine, French chalk, boiled milk, buttermilk, a tomato cut in half or melted tallow. A few of us might remember how to apply a couple of these remedies. Many of us would wonder what on earth French chalk is (it is a solid bar of talc, if that helps clarify things at all). The point is, we no longer have to remember how to apply these remedies. The simple application of a single (complex) detergent will suit all stains.

The detergent may be effective, but the simplicity it represents to the consumer comes at the expense of environmental cost (and environmental complexity) – in the form of modern chemicals that are unlikely to be as biodegradable as more traditional stain treatments. The process of simplification (for the consumer) actually results in a complexity in mixing together many materials (in production).

One consequence of the complexity of fast goods is that the products or the waste from the products can become harder to dispose of in an environmentally friendly way – and it is likely that they are also more resource intensive because (by definition) they are not recycled or recyclable materials.

The problem of complexity and waste is not just an issue for super-fast goods, of course. As we saw in the last chapter, there is an argument for splitting durable goods into their component parts to facilitate recycling. Durable goods and their associated fast goods simplify life for the consumer, but (as with detergents) are more complex in themselves. This whole situation can be seen as a solution to some resource constraints (time or human energy), but the solution itself creates other (environmental) problems.

So, why does the consumer not regress and go back to a less complicated life in the wake of the financial credit crunch? The answer is because the whole point of the complexity of 'needed' fast goods is that, *from the consumer's perspective*, they simplify life. This is, after all, why they are 'needed'. The social

Box 7.2 Volume and waste with 'needed' product

After the Second World War, soaps were rapidly replaced by detergents for use in domestic laundry; they did a better job. By the late 1950s, twice as much detergent as soap was being manufactured in the US. Volumes rose, because detergent was 'needed'. This was a good example of a technological development generating a widespread benefit, in making a labour-intensive job easier to do. Once the consumer embraced detergent, there was no slipping back to soap.

As new technology often does, it brought with it another problem: waste. Older readers may be able to recall exuberant suds forming on the surface waters of sewage treatment plants, sometimes also forming on the more turbulent 'clean' waters of rivers and streams. The cause was a surfactant called alkylbenzene sulfonate (ABS). ABS is not biodegradable, and persisted in treated waters when returned to the environment.

The foaming prompted tests, which found ABB in groundwater. In some cities, beer was not the only liquid that formed a head when poured – a glass of water drawn from the tap did so too. This was when alarm bells began to ring: once groundwater is polluted it can take centuries for unwanted substances to be washed out, and the effects on human health of absorbing even small amounts of ABB over the medium term were unknown. Governments in the UK, Germany and the US responded in various ways, with the common aim of replacing non-biodegradable detergents with biodegradable versions. Detergent manufacturers also responded in various ways and eventually the threat of government regulation triggered the research that led (in the US) to the development of an alternative detergent in the mid-1960s.[2]

What stands out in this particular case is the amount of time it took for a consensus to be reached on regulation. This consensus was ultimately unnecessary, because the threat of tight regulation motivated corporations to innovate. In a nutshell, the commercial sector is where to look for groundbreaking innovation. The challenge is to find the social catalyst.[3]

unacceptability of returning to old ways of doing things is also reinforced. Edwardian advertising promoted the 'feminine virtues of comfort, decency … health and hygiene' in selling soaps and starches, but the image of the laundress was associated with the semantic nexus of dirt, sex, immorality and punishment.

Today, the washing machine that simplifies a consumer's life (though complicates the environmental credit crunch) is an acceptable feature of the modern kitchen. At the start of the 20th century, 'face to face contact between the defiled and the undefiled was to be avoided at all costs' (Sambrook, 1999, p229). Leaving to one side the many possible social and cultural assumptions potentially driving such a comment, the association of laundry with defilement may have had something to do with the inevitably unpleasant smells that would have been associated with lye-based laundry. Almost regardless of the consequences of the financial credit crunch, there is going to be no turning back to a 'simpler' way of doing things when it comes to laundry.

This 'need' for simplicity (from the consumers' perspective) means in turn that the volume of needed goods consumed is unlikely to react to the financial credit crunch. The ratchet effect of simplification is unlikely to be overcome, even by something as severe as the financial credit problems the world currently faces.

The need for fast goods in the wake of the twin credit crunches

The financial credit crunch is unlikely to have a significant impact on demand for fast goods derived from 'need', whether perceived or real. The ratchet effect on demand, and the refusal to degrade one's standard of living, will not reduce the volume of super-fast fast goods. For slower-moving fast goods, there is likely to be an attempt to increase the duration of what one already has. The environmental problems arise in the future, when replacement is needed.

Needs are not the way to change fast good consumption patterns. The environmental credit crunch must look instead to the issue of fashion.

The financial credit crunch and the fashion for fast goods

The Ealing comedy, *The Man in the White Suit,* that was cited in the previous chapter assumed that industry and labour would combine to oppose something that was infinitely durable because, once everyone owned such a product, demand would be sated and economic activity would cease. This seems a somewhat unlikely premise. A white suit of infinite durability, but styled with shoulder pads from the 1980s, would hardly be considered usable today. No one wants to look like a member of the supporting cast of the *Dynasty* soap opera. The vagaries of fashion have a considerable impact on patterns of consumer demand. The more frequently fashions change, the faster the pace of turnover in fast goods.

What is covered by 'fashion' is a fairly flexible concept. In the early 18th century, provincial women were *au fait* with the changing metropolitan fashions for clothes – the publication of annual 'pocket books' (a sort of Filofax for the Georgian gentry) would comment on the latest trends sported by the nobility in London. However, following the shifts in fashion was eschewed. Avoiding the extremes of high society was effectively a defining characteristic of gentility and considered a sign of prudence and good breeding. In other words, eschewing high fashion was genteelly fashionable.

By the end of the 18th century, fashion was clearly becoming more important. What one wore was no longer dictated by social station, but by fashion. The poorest citizens of Paris had clothes that were worth 7.5 per cent of the value of their possessions in 1700, but by 1789 this had risen to 16 per cent.[4] Female servants in 18th century Britain would spend the bulk of their wages (sometimes more than they earned) on clothes. Clothing and other possessions were increasingly considered disposable – fast goods. As time progressed, the value of individual possessions assessed for probate purposes (i.e. after death) declined – because used goods also implied older goods, which were no longer fashionable and thus had less value.

We should not fall into the trap of assuming that fashion is purely an issue that pertains to clothing. Most of the contents of the fast good basket are subject to fashionable forces. Think of the trends in television sets in the past few years. Working televisions were cast aside with casual disdain because they were not flat-screen televisions. The technological advance of flat-screen televisions was largely aesthetic – a matter of fashion and form, rather than function. Mobile phones are essentially considered fashion accessories by the younger generation – to be updated with slightly greater frequency than one's hairstyle.

Clearly, a combination of factors is behind the fashionable demand for fast good items. Communication of fashion is a necessary part of the process – but this is not a sufficient condition (as the early 18th century ladies with their pocket books suggest). Societal pressure is also part of the process – the desire to conform, along with basic good sense, is behind the fact that *Dynasty* shoulder pads are no longer seen on the streets of the world's major cities today. However, the ability to spend money will also dictate the ability to enjoy fashion. This is where the economics of the financial credit crunch has a bearing on fashionable demand for the fast good sector.

The economics

The pursuit of fashion has been described as the 'collective groping for the proximate future' (Blumer, 1969, p281). This groping for the future naturally allies fashion with the availability of financial credit. If financial credit is about having tomorrow's consumption today, fashion is about having tomorrow's style today (or at least, today's style before anyone else). As credit increases, so the ability to sate the desire for fashion will also increase. The intimate interrelationship of credit and fashion was in evidence a quarter millennia ago, with female servants borrowing from their employer in order to purchase 'finery'. It has continued today with the modern accumulation of credit card debt and store credit cards. This means that the financial credit crunch should, in theory, slow the fashion-derived demand for fast goods.

Certainly, the moderation of credit card debt has been a key feature of the financial credit crunch. Credit card debt is expensive in the long term, but readily accessible in the short term. As such, it is ideally suited to short-term purchases (where the borrower has the expectation of repaying the debt in the medium term). It is a form of credit that therefore matches the demand for fashionable fast goods. Credit card borrowing has fallen as the availability of credit has shrunk.

In the US, the credit standards for credit cards have tightened consistently since the end of 2008, even as other forms of credit have seen some moderation of restrictions. UK credit card debt outstanding at the end of 2009 stood 6.3 per cent below the peak of January 2006 – which means of course that the credit card debt to income ratio has fallen more significantly – as banks have tightened credit standards. Borrowers have also reduced their demand for credit card debt it would appear, as a changed view of the economic prospects has inclined consumers to adopt a more conservative budgeting approach.

This does not mean, however, that demand for fast goods has fallen as significantly as the financial credit crunch would suggest. The demands of fashion are a powerful force. The consumer is not confronted with a direct choice between purchasing fast goods and constrained credit. The consumer can also trade down in terms of the quality of the fast good item that they purchase and maintain their sense of fashion (though, perhaps, not their sense of style). If the consumer is prepared to compromise on quality, then the *volume* of consumer goods purchased can remain the same, even as the *value* of the consumer goods purchased falls.

The environment – fashions and fast goods

The way in which fashion and fast goods interact is powerful. Indeed, it is perhaps one of the more powerful forces in modern economics. It also has some potential in terms of the environmental credit crunch. Fashion, if shaped appropriately, can be a force for repairing the damage of the environmental credit crunch – and can push Earth Overshoot Day a little further into the future. However, fashion can also be a negative force in the environmental credit crunch, because fashion (by its very nature) encourages unnecessary change. To understand this, let us turn once again to the different ways in which fast goods impact the environment.

Fast goods, fashion, resource intensity and toxicity

The fact that change in consumption patterns can take place under the influence of consumer aspirations for a better quality of life holds out a great deal of hope for changes that could render fast goods more environmentally friendly. What is needed is for lower resource intensive, environmentally friendly consumption to become high fashion. The question (at least for this book) is whether the financial credit crunch is likely to help or hinder that process.

Palm oil is a good example of a product that has a relatively contentious environmental profile. Its production is alleged to be responsible for the removal of large chunks of the rainforest. At the same time, it is (in a derived sense) fashionable – found in cosmetics, creams, detergents, shampoo and soap as well as many packaged foods.

The extraordinary range of applications for palm oil can be inferred from the range of participants in the Roundtable on Sustainable Palm Oil. This group includes palm oil producers, processors, traders, consumer goods manufacturers, retailers, banks and investors and non-governmental organizations (NGOs). The invisible chair at the table belongs to the consumer, who, in one sense, potentially has a great deal of power (imagine the effect of a boycott of products containing palm oil). In another sense, the consumer has absolutely no power at all, often having no access to information about which oils are used in any given product (let alone their environmental credentials).

How can consumers force change so as to allow palm oil to be used (as a substance needed in many fast goods) without damaging the environment? Fashion plays a role here. If a product can claim to be using palm oil derived in a suitably environmentally friendly (and sustainable) manner, then that may have a fashionable cachet.[5] Many fast good manufacturers depend on the perceived value of their brand to achieve commercial success: the brand must be fashionable. If the consumer responds positively to products using environmentally sustainable oil (i.e. the consumer is prepared to pay) or negatively to products using oil that is not perceived as environmentally sustainable (i.e. a consumer boycott), then the fast good manufacturer has a powerful incentive to respond.

So what does the financial credit crunch mean here? Essentially, the incentive structure becomes lopsided. With credit constraints, the consumer is less likely to be swayed by fashion into paying a premium for environmentally friendlier products. Let us assume, in the palm oil example, that cosmetics with sustainable palm oil sourcing require a 10 per cent premium. With a wide range of products on the market, a certain amount of inertia as to brand selection and a more constrained financial credit environment, the consumer is unlikely to switch to the product that helps to resolve the environmental credit crunch.

On the other hand, the boycott argument still works. If peer pressure can suggest that it is not fashionable to purchase cosmetics using unsustainable palm oil, then the demand shift can be accomplished against the backdrop of the financial credit crunch. Indeed, if the fast good is being demanded more because of fashion than because of need, this is even more likely to happen. In this case the changing dictates of fashion coupled with a financial credit crunch will actually prove to be an incentive to reduce consumption of the product thereby helping to reduce the environmental credit crunch.

Fast goods, fashion, volume and waste

When it comes to fashion and fast goods, volume and waste go hand in hand. To throw something out because it is 'unfashionable' is to dispose of something before its practical life is over. The volume of fashionable demand must therefore be closely associated with waste of resources and the two can be considered to be most intimately intertwined.

The supply-side issue is whether the supplier of the product can render it more environmentally friendly while still maintaining the balance between costs and a reasonable price in the market. The demand-side question is whether the bundle of influences driving consumer need and desire for certain products could be reshaped.

Box. 7.3 Fashion fighting waste – gift-wrapping the environment

The philosophy behind 'make do and mend' – the use of fast goods as often as possible and in as many different ways as possible before disposing of them – used to apply to the packaging of fast goods. With the advent of plastic, the practice of taking empty bottles back to the retailer for recycling almost disappeared.[6] Many fast good items are either made of plastic or come so profoundly sealed up in it that the skills of a safe-breaker are required to access them.

However frustrating this is there are reasons for such practices relating to the need for the manufacturer to maintain product integrity. Due to such perceived benefits, a move to 100 per cent recyclable packaging is unlikely.

Packaging is one area where fashion is changing consumer behaviour and thus the environmental credit crunch. There has been some effort to reduce the impact of packaging – beyond the now common recycling of cardboard. Detergent manufacturers have moved towards concentrated product – which involves less packaging (and, of course, reduces transport costs in both an economic and an environmental sense). With the consumer accepting that concentration is not really a problem for a product that by definition is diluted in its use, waste is reduced. Supermarkets charging for plastic bags and encouraging the use of reusable bags is further evidence of the ability of fashion to change waste. When the supermarket Sainsbury's launched the 'I'm not a plastic bag' cotton shopping bag in 2007, stores sold out within an hour. Environmentalism became a fashion statement (it helped that the bag was used at the 2007 *Vanity Fair* Oscar Night party). Marks & Spencer has reduced plastic bag use by 64 per cent from 2006/2007, reducing both volume (of bags produced) and waste (disposing of plastic bags).

The impact of the financial credit crunch here is mixed. In one sense, cost-sensitive consumers are likely to respond well to the (economic) cost savings of reduced packaging. The demand side is encouraged by the financial credit crunch. The problem lies with the supply side. Where companies use packaging to protect product integrity, reducing packaging will require more research and development (R&D). R&D is likely to be less readily forthcoming in a financial credit crunch world (because it is a form of investment, which is now more expensive to undertake).

Fashion and fast goods in the wake of the twin credit crunches

Fashion can play an important role in reducing waste, as the plastic bag example (see Box 7.3) makes clear. However, the ability of the supply side to meet those demands is somewhat constrained by the financial credit crunch. Of course, if the demand is strong enough that consumers are prepared to pay a premium for the fashion of environmentalism, the supply side has an incentive to pay for the research required to achieve those aims, in spite of the constraints a more economically austere environment imposes.

If, however, the consumer is not prepared to pay a significant premium for environmentally friendly fast goods (because of the financial credit crunch) but only to boycott those that are not seen as environmentally friendly, there is a complication in handling the environmental credit crunch. Without the prospect of a premium, the supply side has to face increased R&D costs with no certainty of reward, which, at best, will complicate attempts to push out Earth Overshoot Day.

The financial credit crunch, envy and fast goods

In a materialistic world, material possessions can be thought of as a defining standard for social status. Of course, durable goods will be part of the standard, but the fast good sector is the more readily accessible (because, generally, it is more affordable). Aspiration or envy is a clear part of selling the fast good sector – and something that advertisers spend a great deal of time and effort to foster. Of course, envy is something that overlaps with the other forces of demand for fast goods: advertisers can try and teach the consumer that they 'need' something, by presenting the standard of living associated with ownership as the minimum acceptable; similarly, deriding something as unfashionable while simultaneously presenting an alternative as a worthy ambition for the fashion conscious may also create demand from aspiration and envy.

IKEA's 'Chuck out your chintz' advertising campaign from the turn of the century was about fashion, but also about trying to persuade consumers to *aspire* to a specific lifestyle. To the extent that this could have involved the destruction of serviceable furnishings, it could be considered as potentially productive of environmental waste, unless something constructive such as recycling were proposed for the unwanted chintz. It is unlikely the company was deliberately seeking to portray a consumer durable good as a fast good, thereby encouraging consumers to embrace waste, but the consequences of consumer behaviour potentially encouraged by the message embedded in these adverts must be confronted.

Using envy to sell is hardly a new concept. Over half a century ago it was outlined with rather chilling clarity by an economist (inevitably), Victor Lebow, who wrote: 'Our enormously productive economy demands that we make consumption our way of life, that we seek our spiritual satisfactions, our ego satisfactions, in consumption. The measure of social status, of social acceptance, of prestige, is now to be found in our consumptive patterns.' This analysis was contained in the Spring 1955 edition of the *Journal of Retailing* – and what reader has not browsed through the contents of the *Journal of Retailing* on some rainy Sunday afternoon?

The economics

The reaction of the fast good sector to the financial credit crunch is somewhat perverse. Generally speaking, fast goods are a relatively small part of consumer spending.

In *The Road to Wigan Pier* (published in 1937), George Orwell identified the advantage of cheap goods that the wider population could aspire to: 'The thing [demand for fast goods] has happened, but by an un-conscious process – the quite natural interaction between the manufacturer's need for a market and the need of half-starved people for cheap palliatives.' This 'cheap palliative' demand is not fast good items determined by need. It was aspiration and gratification combined in a heady cocktail. Orwell went on to describe the attitude of the unemployed in 1930s Britain to food:

> *The ordinary human being would sooner starve than live on brown bread and raw carrots ... When you are unemployed, which is to say when you are underfed, harassed, bored, and miserable, you don't want to eat dull wholesome food. You want something a little bit 'tasty'. There is always some cheaply pleasant thing to tempt you.*

Orwell is talking of food, obviously, but the concept of being rewarded by something a little bit 'tasty' is just as prevalent in the fast good sector today.

Think of the Apple iPad. (At this point in the proceedings the other writer was tempted to take on the identity of Marvin the robot in *The Hitchhiker's Guide to the Galaxy* – 'Oh, go on then.') In the midst of a financial credit crunch, no one really *needs* an iPad – at least, certainly not in the quantities that are demanded. (Here, Marvin would grudgingly agree.) An iPad does not, generally, have any advantage over any laptop in terms of its functionality. (Here, Marvin would object that the iPad app is a game-changing piece of technology.) Why should a laptop owner purchase an iPad at all, assuming their computer continues to function? (Marvin would observe that this is obvious to anyone who has used both.) Fashion could play a role here, but it is stretching things a little. Go through any airport security screening process and it is evident that the appeal of the laptop has not diminished in any notable way. (Marvin would point out that we are all stuck with our laptops because the credit crunch has affected our ability to afford the switch to iPads.) No, the demand for the iPad must be considered to reflect the aspiration/

envy-induced demand of the fast good sector. 'They have one, so I want one,' combined perhaps with a sense of entitlement. 'I have been good, I have paid down my credit card bill, and *now I deserve a treat*.'

As Marvin's slightly irreverent intervention serves to remind us, it is difficult to separate need from fashion from improvements in the quality of life, and when new technology arrives, any environmental gains or losses tend to be accidental or undefined. The concept of envy is hardly new (the seven deadly sins are not exactly a modern concept, and the Old Testament of the Bible warns against 'coveting thy neighbours' goods'). Today, it is iPads and similar electronic wizardry. In the early 18th century, the pocket watch was the iPad of its day. The pocket watch was invented in the late 17th century, but within 100 years watches were owned by nearly 40 per cent of paupers in the UK. A pocket watch cost several week's wages, but it was a symbol of status and something that the very lowest in society would aspire to. The potter Josiah Wedgwood understood this more than most. He sought to appeal to snobbery and what has been called 'nobility envy', to sell what basically amounted to cups and saucers. In doing so, he laid many of the foundations of the modern retail society.

William Makepeace Thackeray's *The Book of Snobs* (published in 1848) covered snobbery and the concept of envy or aspirational demand. Tellingly, Thackeray sought to elaborate what a snob was in the book, because the term was still relatively novel. It was born out of the materialism and faster fast good environment of early 19th century society.

Here we see the perversion of a financial credit crunch on aspirational demand for fast goods. If consumers are constrained as to income and credit, but not so constrained as to lose all flexibility, the demand for some fast goods will tend to substitute for other forms of demand as consumers seek to compensate for the loss of that 'feel good' factor that tends to go with consumption. Consumers will thus economize on their food spending. They may also forgo (because they are forced to forgo) spending on consumer durable goods. However, if they have successfully managed these privations they will 'reward' themselves with a treat. The fast good sector provides an abundance of such treats, falling within an affordable price range.

The environment

Envy is hardly a noble motive for anything, least of all for consumer spending. Purchasing goods out of envy, regardless of need, is likely to entail environmental waste because the old possessions must then be disposed of. Even if the regulation discussed in the consumer durables sector is a move to recycle at least some of the electronics, what this chapter suggests is that the technology-switching cycle may be altogether too rapid from the perspective of the environment. To consume finite resources purely to gratify one's aspirations could be seen as environmentally wrong. The problem is that to stand against the gratification of envious desires is to be labelled a 'killjoy', and of course environmental opposition to consumerism in this form is definitively killing the joy the consumer wishes to feel.

However, there is a silver lining to this particular cloud, already hinted at in the earlier sections, which noted that environmental impact in the context of consumer goods tends to be left to chance, noting that chance can cut both ways. In the context of slower-moving fast goods (or should that read faster-moving durables?) it is possible that fast good envy will produce a more (environmentally) efficient outcome, albeit by chance. Let us consider the provision of music and entertainment. Someone who owns a CD player is already satisfying their 'need' to hear music. Need plays no motive in shifting demand from a CD player to an MP3 player. Fashion may play some small part in demanding an MP3, but the role of fashion is more likely to dictate the brand of MP3 player than the transition from CD player to MP3 (the continued existence of CD players, and the fact that MP3 music systems allow their owners to burn their own CDs, would argue that this is not an entirely fashionable transition). Envy and aspiration must play a role in the desire to make the change.

So far, the action of the consumer is environmentally reprehensible. A perfectly satisfactory method of meeting a need has been cast aside on grounds of envy or aspiration, perhaps overlaid with some concerns about fashion and appearance. However, the shift also represents a change in technology that is, in fact, economically beneficial.

By switching from a CD-based music delivery system to an MP3 player-based system, the consumer is significantly altering their environmental impact. The consumer is no longer purchasing a physical CD, with associated packaging (including a small plastic carrier bag). Instead, they are purchasing a stream of electronic information downloaded legally (one trusts) from the internet. This is not a costless process in environmental terms, however. Computers (and the servers that support the electronic music store) consume energy and create carbon footprints – but overall (considering the point that an entire MP3 player can weigh just one-third of a single CD and its plastic case), the process will generate fewer lost resources than accumulating a large collection of CDs.

What is happening here is a happy coincidence between the economics of envy and the needs of the environment. The financial credit crunch does not reduce the aspiration of the consumer and, in the sphere of fast goods, may actually increase it. However, if that aspiration for ownership combines with a more environmentally efficient pattern of consumption, then the environmental credit crunch can be lessened. Buying an MP3 player, even if it is not needed, may help move Earth Overshoot Day a little further away.

Of course, not all aspirational desires will be satiated by environmentally efficient products. Aspiring to a new pair of shoes or a designer coat is not creating a paradigm shift in terms of environmental consumption trends. The outpouring of plastic that is the children's toy industry is not generally speaking creating a more environmentally efficient world (and was there ever an industry more dependent on envy as a source of demand than that of children's toys?). The trick for society is to make environmental efficiency an object of envy or aspiration. If it can succeed in this, then the shifting patterns created by the financial credit crunch could be environmentally beneficial.

Conclusion: Fast goods and the twin credit crunches

The financial credit crunch impacts the fast good sector in diverse ways. Where there is a need, there is hope that demand patterns will seek to increase the duration of fast goods, and lessen the 'fast' aspect of the broad fast good definition. Using computers for longer than would have been the case prior to the financial credit crunch and adopting a 'make do and mend' approach – these are common approaches.

However, the risk is that when the repairs are no longer economic, the replacement is less durable. This also applies to some extent to fashion. Tastes and trends change with the same frequency in a financial credit crunch as in an economic boom. Although the urgency of fashion may fade somewhat, it is unlikely to make a material difference to the desire to own the latest object – or to the desire to cast aside last week's fad if it no longer suits the current style. In short, a shift towards a 'culture' of frugality as a consequence of the credit crunch could help reduce the over-consumption that currently plagues a number of economies, but it is unlikely to be sufficient in isolation to inculcate a sustainable approach to the consumption of staples.

Even if frugality on the part of consumers as a group could mitigate the fast good volume problem, it cannot directly affect what goes into the product, nor can it directly affect product design (which often determines frequency of use and replacement). However, the consumer companies that make fast goods have to compete hard to stay ahead of their competition and do so by meeting the demands of their customers so, indirectly, the consumer has significant power to change the environmental footprint of fast goods. At the margin, the aftermath of the crunch could help bring about a shift in the consumer mindset, but it is unlikely to be enough without a broader cultural change. The fact is, environmental considerations need to become so much a part of branding that designing a new product without taking the environment into account, or indeed purchasing a new fast good without doing the same thing, becomes reprehensible.

Notes

1 This measures how intensive the use of environmental resources is, to produce, distribute and consume a fast good product. One way to think about this is in terms of

the ecological footprint of the product, measured perhaps in the square metres of land that would be required to supply the ecosystem services (water, raw materials, greenhouse gas (GHG) sequestration and so on) needed to manufacture one tonne of product, across the *entire* cycle of production and usage. The lifecycle analysis introduced in the chapter on durable goods (Chapter 6) is a similar idea.

2 The research produced the so-called 'molecular sieve' technology that led to the development of linear alkyl sulfonate (LAS).

3 The importance of the strong social catalyst is not to be underestimated when fast-moving goods are widely used. Another topical example is found in the presence of phosphates in detergents and shampoos. Phosphates are not toxic, in fact they are a necessity to living organisms, but released into water courses in sufficient quantities they upset the balance of nutrients so that algae thrive to the detriment of other species. Phosphate-free detergents – another product of successful R&D – have been available for almost two decades, but the catalyst required to bring about their widespread usage is still awaited.

4 The details of clothing inventory come from De Vries (2008, p139). See also De Vries (2008, pp124, 143) for the broadening of rural workers' inventory of possessions, and the prevalent use of credit by female servants to purchase clothing.

5 GreenPalm is one method. The process works in a similar way to the better known concept of carbon trading. Producers of certified 'green' palm oil are able to sell a certificate (once) to that effect to palm oil buyers. The buyer can claim to be using (more) environmentally friendly palm oil, even if the actual source they use is not 'green'. The point is that the consumer knows that by buying a product containing certified palm oil, an equivalent amount of oil has been produced in an environmentally sustainable manner *somewhere* on the planet.

6 Recycling glass at a bottle bank is a poor substitute – because energy is required in reprocessing the glass. The traditional method of returning the bottle was less energy intensive as the bottle could be reused in its existing form.

References

Blumer, H. (1969) 'Fashion: From class differentiation to collective selection', *Sociological Quarterly*, vol 10, pp275–291

De Vries, J. (2008) *The Industrious Revolution*, Cambridge University Press, New York

Sambrook, P. (1999) *The Country House Servant*, Sutton Publishing Limited, Thrupp, Stroud

CHAPTER 8

The Twin Credit Crunches and Human Health

Bona valetudo melior est quam maximae divitiae (good health is worth more than the greatest wealth) – Latin proverb

The previous chapters of this book have set out a variety of ways in which consumer and economic behaviour has been altered by the financial credit crunch, and how this then in turn impacts the environmental credit crunch. For the consumer, the main impact of the financial credit crunch is that a fall in income or credit (or the fear thereof) can change consumption patterns (the mix of goods and services income is spent on) as well as the absolute amount spent in any given year. The consequences of these changing consumption patterns are (we hope it is clear by now) diverse in their environmental implications.

The relationship between health and the twin credit crunches is more complex. What we actually have is a 'ménage à trois'; how environmental credit crunches affect human health; how financial credit crunches affect environmental credit crunches, and through that health; and how financial credit crunches affect health directly.

These three issues have a triangular web of interrelationships, which is what this chapter sets out to explore. Before we roll our sleeves up to examine the situation more deeply, however, we need to consider the particular characteristics of health-care related spending, for it is not so easily confined or compartmentalized as some of the other sections of this book. The economics of human health can be narrowly defined as the formal health-care sector. It can also be defined far more broadly as 'human welfare'. That includes the full range of human needs which, if met, will improve general welfare and quality of life, including health. We are considering health in a fairly wide sense here. This is more than how hospitals fair in the face of the twin credit crunches. 'Health' in this sense moves beyond pharmaceuticals and surgeons, and into the more general realm of well-being or *bien être*.

The diverse range of inputs into health can be demonstrated by considering health and welfare in developing countries. The United Nations (UN) Millennium Development Goals encompass a number of themes which connect to health. Some goals – to reduce child mortality, improve maternal health and combat disease – are obviously directly health related. Others goals are likely to have indirect impacts on health: goal seven, 'Ensure Environmental Sustainability', includes the provision of clean water and sanitation. Experience has historically shown this has a huge impact upon health. Goals five and six are concerned with raising educational standards and empowering women, and are likely to have positive indirect impacts upon health through educating consumers to take care of their health.

The first Millennium Development Goal – the eradication of poverty and hunger – is of course likely to have the greatest impact on the health of the population through raising general standards of living. However, raising those standards of living requires economic development and has environmental consequences. Thus, the financial and environmental credit crunches become intimately intertwined with health.

There are many possible approaches to this subject. We have focused on four strands. Although these four strands are treated as distinct, in an attempt to give a degree of clarity, there are of course many feedback loops across the different concepts. What this chapter examines is the following:

1 The way the environment impacts health (and how the financial credit crunch will impact the environment).
2 The overconsumption of resources and the implications for health. We also here take the perhaps somewhat 'risqué' approach of suggesting that health care is a resource that can be overconsumed.
3 The way the financial credit crunch impacts working patterns and behaviour, and through that impacts health, either directly or indirectly.
4 The way the financial credit crunch impacts spending on prevention, cure and research and development (R&D), and what that implies for health.

This is not the straight-line approach of previous chapters. We cannot go from financial credit crunch to altered consumer behaviour and thence to the consequences for the environmental credit crunch. The consequences for the environmental credit crunch here are not necessarily the outcome (as it has been with other chapters). For health, the environment is more likely to be a staging post en route to the health-care conclusion.

From the environment to health

The constant theme of this book has been how the financial credit crunch will shape the environmental credit crunch. After seven chapters of admirable adherence to this causality, why should we now choose to tamper with this flow? The answer is back with the 'ménage à trois'. The environmental credit crunch impacts health and, through health, impacts the economy. The causality surrounding the twin credit crunches is far more of a two-way process here.

Let us use an historical example to illustrate the point. The 'pea souper' smog in London during the 1950s was the consequence of economic activity (as coal smoke from urbanization and industrial activity combined with local weather conditions). It created a physical environment that was detrimental to health. Indeed, estimates attribute up to 12,000 premature deaths to the smog. Ultimately, there is an economic feedback (the ensuing ill health and fatalities have an economic as well as a human cost). Going to first causes, one can find economics at the root of the problem. The immediate manifestation is an environmental problem. The consequence is a health problem, which leads to economic problems. The solution lies in legislation that aims to break the chain of events, not with the economics (obviously) but with the environmental consequences of the economic position.

From the financial credit crunch to the environmental credit crunch, and so to health

As we have seen in the course of this book, the financial credit crunch has a very varied impact on the environment. Growth is likely to slow, which means that the pace of damage to the environment should also slow (though not cease, far less reverse). However, investment spending is also likely to slow, which presents problems in upgrading infrastructure, with environmental consequences. Consumption patterns shift in ways that may or may not be environmentally efficient.

What the financial credit crunch is doing, therefore, is instituting change which will alter the pace of change in the environment. The issue at hand is how that change in the environment will have a bearing on health.

Is the environmental credit crunch bad for your health?

The UK's Environment Agency states: 'A high-quality environment enables people to live longer in good health.' On its official website, the World Health Organization (WHO) defines environmental health as 'all the physical, chemical and biological factors external to a person … that can potentially affect health'.

According to the WHO, one-third of all death and disease in developing countries arises from environmental issues. In developed countries, preventable death and disease attributable to environmental problems includes cancer, cardiovascular disease, asthma, lower respiratory infections, musculoskeletal disorders, as well as problems that seem to have come from the traditional whodunnit – injuries, poisonings and drownings.

Longer life expectancy is generally associated with a healthy environment because, at a basic level of economic development, life expectancy is influenced by nutrition, clean water supplies and clean air among other things, and not simply by the availability or otherwise of health services. In wealthier countries (where decent standards of environmental management and food provision tend to be taken as given), it is also explained by differences in society, diet and how health care is provided. These indicators can explain further differences in life expectancy from one region to the next (see Smith, 2008, p20).

Thus, as economies become more developed, the impact of the environment on health tends to become increasingly indirect. The effect of the environment on health nonetheless remains powerful in developed economies, because if environmental maintenance weakens for any reason (including budget constraints after an economic downturn)[1], direct effects (such as those arising from industrial accidents or from the difficulty of containing the environmental impacts of large amounts of economic activity) can become relevant.

Environmental credit crunches can seriously damage your health

If the environment is unhealthy, it is unlikely the human beings who depend on it will be healthy, either. The Earth Overshoot Day concept that runs through this book suggests that the environment may be moderately overloaded and thus suffering from ill health. Even this concept might mask the severity of the health problem for some. Earth Overshoot Day is an aggregation of environmental demands. Some areas will undoubtedly suffer greater localized environmental concerns than those implied by Earth Overshoot Day.

The London 'pea soupers' of the 1950s may have been consigned to history, but city-specific environmental problems have hardly disappeared. In China, as the 2008 Beijing Olympics approached, car travel in the city was restricted to try to clear the air of the worst effects of air pollution; some marathon runners took the decision not to participate for the good of their health. This is not just an emerging market issue. In the US, a recent government report commented: 'Mobile-source air emissions, especially diesel particulate pollution, are responsible for approximately 30% of cancer resulting from air pollution.'[2]

In Budapest, in October 2010, the wall of an alumina tailings reservoir broke, spewing 184 million gallons of toxic sludge onto three villages. The water system around the area of the spill will clearly take many years to recover (if it ever does). The spill got as far as the Danube but by that time seemed to have become diluted enough for pH levels to be within a normal range. Flushing away, neutralizing with other chemicals, dispersing by mechanical or chemical means, are the usual responses to such spills. What is forgotten is that some unwanted chemicals will remain permanently in the environment, if they do not decompose or degrade to the point of becoming chemically inert. There is ultimately no such thing as flushing 'away'.

The relevant unknown is what the long-term health effects will be from the heavy metals that must (inevitably) end up in the water table – and also perhaps in the raw materials that go into food. The potentially very long-term environmental impacts of human activity are also something that is often forgotten.

In Johannesburg, one legacy of more than a century of gold mining is a rising pool of acidic mine water. There are fears that, without remedy, this acidic water could reach the streets by 2012. As human populations grow and it becomes harder and harder to maintain a good balance between human systems and the ecosystem, the risk of such pollution 'hot spots' can only be rising. At best, they might render cities less healthy and, at worst, render them uninhabitable.

Urban environment

The urban environment is a 'synthetic' environment (rather than the natural environment under discussion in this book) but deserves mention in this section on health because it is (along with some modern health problems) a product of modern lifestyles. Most people of a certain age in the West probably recognize intuitively that asthma seems to be more prevalent than it was in their youth. It is reportedly the most common chronic condition in Western countries and, while the causes are still not fully understood, the list of suspects includes air pollution (noting that indoor air can contain pollutants such as fumes from cleaning materials and heating systems) and centrally heated homes (associated with a rise in dust mites), as well as processed food.[3]

Environmental factors can also be responsible for raising cardiovascular risk – for instance, in the aged it can rise significantly in the presence of indoor pollutants. Although the urban environment is generally designed to be as safe as possible (in developed countries with regulation in place, at least), such unexpected effects can affect the more vulnerable residents. Other examples include the so-called 'heat island effect' (concrete acting as a storage heater in hot summers, to the discomfort of those living within) that hit Chicago in 1995 and Paris in 2003, with a few thousand people dying from the effects of the heat. Some of the problems affecting the urban environment are a product of that specific synthetic environment but others (such as the heat island effect) are likely to be a problem in the context of climate change.

Beyond the cities – wider environmental problems and health

Perhaps the most widely discussed consequence of environmental degradation is climate change. The *Stern Review* observes that climate change 'will alter the distribution and incidence of climate-related health impacts', reducing the number of cold-related illnesses and death, but increasing the effects of heat stress, drought and flood. In developing countries, drought also tends to be associated with increased malnutrition – 70 per cent of food emergencies are caused by drought (Smith, 2008, p110) – and gastroenteritis. The spread of vector-borne diseases such as malaria is also likely to change with global changes in temperature.

Health and the environment

What these examples are illustrating is the very clear link between the environment and health. There is a common thread that runs through these examples. The environmental changes that are producing health consequences have economics at their heart. This does not mean that economics is necessarily going to provide the solution to the health problems – as with the UK's response to the London smog (the Clean Air Act), it may well be that the health solution lies with better environmental management rather than with a change in the economy per se. This suggests that the direct links from financial credit crunch to environment to health are not going to be terribly robust.

However, this is not the only mechanism through which the twin credit crunches impact health. The issue of overconsumption has a very direct bearing on health and this is where we now turn our attention.

Overconsumption and health

The importance of food to health should be obvious. Health, even broadly defined, is essentially a measure of how the human engine is running. Food is the principle fuel of the human engine (there is water, sunlight and so on, but food is pretty much the main driver). Anyone with even a cursory knowledge of computer science is familiar with the concept of 'garbage in, garbage out'. If the inputs are not right, then the output (in this case, health) is unlikely to be right.

Overconsumption of food creates health problems that are difficult to counteract. The exercise machines of many health clubs helpfully tally up an approximation of the calories burned by half an hour's vigorous exercise. It rarely amounts to more than a chocolate bar's worth (and is often little more than the cork of a bottle of wine).

A better diet lengthens life expectancy and improves health as economies develop. However, too much of a good thing (and too much of less good but affordable things such as fat, sugar and salt) has also led to an increase in the numbers of overweight and obese people in wealthier countries. Using a body mass index measure, more than 70 per cent of Americans are defined as overweight or obese. In several Latin American countries, as well as the UK, Germany and Spain, 50–69 per cent of people are at least 'overweight'. This state of affairs can have practical consequences. In the US, CBS reported that 27 per cent of adults aged between 17 and 24 were too fat to meet army recruitment health standards.

The health problems arising from obesity are legion. It is widely accepted that being overweight or obese increases the risk of stroke, coronary heart disease, Type II diabetes and cancer. Obesity is also associated with less serious problems (from back pain to psychological issues). While these diseases are not strictly speaking environmental, they arise from behaviour that both shapes and is shaped by the environment.

Cheers – good health and the economics of the financial credit crunch

The direct impact of the financial credit crunch and its attendant economic slowdown is an uneven balance of positive and negative impulses on health. As we discussed in the first chapter, food consumption patterns will change as perceptions about income level and affluence alter. However, one of the main shifts in food purchase was a reduction in waste – which means, of course, that the calorie intake of any individual is unchanged (it is simply more efficiently catered to).

Budget constraints also tend to reduce alcohol consumption overall (and will therefore reduce calorie consumption). Growth in alcohol consumption shows a strong positive correlation with gross domestic product (GDP). Consumers do not drown the sorrows of a weak economy with alcohol; they confront the situation with sobriety.

The question is how enduring these changes may be. Some of the changes in consumption patterns may last and the attendant health benefits can be thought to endure. The danger, as always, is that one will slide into a behaviour pattern that 'rewards' austerity in other aspects of the budget with 'treats' that are relatively inexpensive to purchase. Buying a bottle of wine to celebrate the fact that one has met one's savings target for the week, or indulging in chocolate (purchased with cash) to compensate for having resisted a purchase that would have required borrowing on a credit card – these are the trends that can easily emerge.

An example of this inconstancy in the direct health benefit comes from the UK's wartime experience. Wartime food rationing created a diet that was healthy – far healthier, indeed, than the diet of many in the Organisation for Economic Co-operation and Development (OECD) economies today. However, when sweet rationing ended in 1953, the response was not a modest increase in consumption, shaped by appetites that had been restrained by years of limitation. The response was an orgy of consumption that strained the newly formed National Health Service (NHS) with the scale of the tooth decay that ensued. Sweet consumption increased every year until 2001 (see Box 8.1).

Even if transitory, there are some implications from the financial credit crunch for food and, through that, for health. A parallel can be found in China, where the increase in meat prices in 2008 had a direct consequence for meat consumption. The higher cost of pork (the result of a disease) led to reduced consumption of the meat. This is similar to operating under a budget

Box 8.1 Budget constraints and health

The seemingly inexorable rise of confectionery consumption in the UK was halted in 2001. The decline in 2001 was due to the rise of mobile phones. Children were faced with the choice of spending pocket money on chocolate and sweets, or spending money on 'top-up' cards to enable them to send yet more text messages on their mobile phones. It was a contest that even the confectionery industry could not hope to win. Sweet consumption declined, no doubt helped by the fact that UK newsagents tend to sell mobile phone top-up cards along with confectionery. The two products were competing head-to-head. This is perhaps an early and rare instance of consumer choice voluntarily taking the healthy option when faced by constrained income.

constraint (essentially because a consumer with a fixed income is able to buy fewer pork chops on their budget as the price of pork rises). There is a *chance* that the financial credit crunch will act as a constraint on the overconsumption of food, therefore.

The environmental issues of the obesity epidemic

The description of obesity as an 'epidemic' raises a valid question; could obesity reverse the trend of rising life expectancy? Russia has reversed this trend and now has declining life expectancy (a switch that is generally attributed to alcohol consumption). Perverse though this may seem, shorter lifespans may alleviate environmental stresses and perhaps reduce demand pressures on health-care services.

While a fall in longevity could (from some perspectives) be considered a 'convenient' solution to a significant budgetary problem, few would regard it as a 'nice problem to have'. A decline in life expectancy is also not certain. The Foresight Report (applied only to the UK) suggests advances in treating diseases are reducing (and will continue to reduce) their morbidity. However, the financial consequences of this trend – which suggests an increasing treatment burden – are uncertain. Moreover, this trend is also likely to increase the cost of maintaining a decent quality of life for people who become ill beyond their middle years:

> *Poor lifestyle trends such as increasing obesity among young adults and children may in the long term reduce the increase in life expectancy, but over the next 30 years they could make the burden on the working population worse, since they may reduce the number of healthy working-age people more than they reduce the number of elderly pensioners.* (Pensions Commission, 2005)

Health, longevity and the environment

These days, they say 60 is the new 40, and the traditional retirement age of 65 (which used to be followed by death perhaps a couple of years later) can be followed by a 'second youth', assuming those concerned are fit enough to take up snowboarding or whatever their long-hidden passion may be. If poor lifestyle trends are likely to compromise the longevity and by implication long-term health of today's young, then today's middle-aged would be well-advised to think about doing all they can to remain spry well into old age. The reason is that this youthful middle-aged generation is sometimes described as the 'sandwich' generation (caring for aging parents and children at the same time in the early years of the 21st century), while, in the future, there may well be a smorgasbord (open sandwich) generation, with today's young unable in early old age to look after their even older parents.

It is worth contemplating the prospects of those who might suffer the double cost of having cared for two generations in their middle years, followed by an old age looking after two generations all over again (this time around, needing to care for themselves and their now ailing offspring). This still hypothetical situation, should it arise, would be compounded by uncertainty about investment in health services (with potential decade-long effects on future services) as well as uncertainty over the value of pensions (a question mark for the funding of consumer spending on health care in the long run). Once again, although these potential problems are not directly environmental, they are associated with behaviours that are bad for the environment. Addressing environmentally problematic behaviour (which also happens to be bad for the health of several generations) could potentially help mitigate the social and financial health-cost time bomb such as that described above.

Overconsumption of health

The overconsumption of food is something that has self-evident health consequences – as well as forcing Earth Overshoot Day to occur earlier in the year than is necessary. However, health (more accurately, health care) is also a resource-intensive form of consumption that consumers may purchase too much of. Surprising though it may seem, unnecessary spending on health care may also bring Earth Overshoot Day forwards in a completely unnecessary manner.

Whether health care is worth the money spent on it depends on the efficiency with which the money is spent. Health-care economics measures the cost-effectiveness of health-care services in terms of 'cost per quality-adjusted life-year'. At lower levels of spending, the gains from added spending tend to be significant under this measure. Once spending has reached a certain level, however, the benefit levels off.

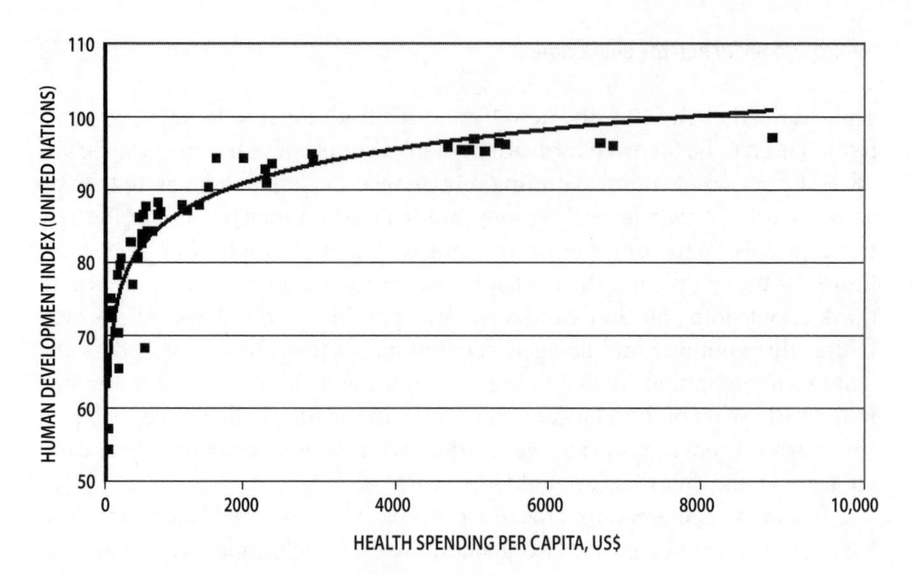

FIGURE 8.1 Health-care efficiency levels off

Source: Human Development Report 2010, United Nations Development Programme (UNDP).

The briefest of comparisons between the US and Europe shows this to be the case. Europe is a society that is ageing a great deal more rapidly than the US is. However, the economic consequences of the two regions' demographics are about the same. The reason is that it is about three times as expensive to be elderly in the US than it is in Europe. This difference in cost is largely due to the expense of US health care. The cost of US health care means it requires roughly thrice the economic resource allocated to health care in Europe. The health consequence is no different, however.

What this suggests is that efficiency in spending money on health is critical to keeping health-care costs down. Spending on illness caused by diet or lifestyle can be viewed as a misallocation because such illnesses could be dealt with far more cheaply by being prevented and therefore dealt with ('treated') outside the health-care system. No matter how good the health-care system, lifestyle and diet can be expected to be more important than the system in determining overall health levels.

What has this to do with the environment? There are two environmental issues. First, and most directly, money wasted represents a resource that is wasted. If energy, expertise or pharmacology is being 'unnecessarily' devoted to health, that is a potential environmental resource that is being 'wasted'.

Second, every 'unnecessary' dollar spent on avoidable health problems anywhere in the world could be regarded as a dollar that could be made available for unmet medical needs somewhere else – whether in providing needed access to medicines and services (within budget) in developed countries or in developing countries. Treating the environment with respect could be regarded as an inherent part of creating optimal conditions for the health-care sector to do its job. In other words, the environmental credit crunch again needs to be considered as an input to health (and the economy), as well as a consequence of health and the economy.

Overconsumption and health in the era of the twin credit crunches

Overconsumption (in the sense of 'excessive' consumption) is something that is by definition detrimental to the environmental credit crunch. However, overconsumption can also be detrimental to health. Health spending is also a potential cause of environmental overconsumption. We have come back to the complexity of the web of relationships when it comes to health, the environment and the financial credit crunch. The next area of focus also exhibits complexity. It is time to look at the way we live now.

The way we live now – the financial credit crunch, work, health and the environment

The financial credit crunch has a wide range of implications for how people live their lives and how people work and, through those channels of influence, how people are likely to perceive their well-being. This broad sense of health will be intimately bound up in the environment in which we live.

The direct consequences of the financial credit crunch on health

A financial credit crunch will always tend to involve change in society. Some people cope with change well, some people cope with change badly, and some people do not cope with change at all. It has been known for over a century that suicides will tend to rise during periods of economic weakness.[4] Mental health problems are known to rise among those who experience long-term unemployment, and the financial credit crunch is certainly a contributory factor here. The recession of the early 1980s, which was a prolonged period of economic weakness (and falling living standards for many in society), saw the unemployed report increased stress, mental health problems and increased use of medication. This was particularly the case, as one might imagine, for the long-term unemployed. One of the consequences of the current downturn is that long-term unemployment has risen (and part of the increase in unemployment may be structural – that is to say, the unemployment becomes persistent).

Apart from suicides, however, the direct negative consequences of an economic downturn on health via stress are less in evidence. Analysis from Europe has shown that other stress-related health issues (heart problems and the like) are not, in fact, particularly well correlated with periods of economic weakness or rising unemployment. Although the extreme of suicide is clearly correlated, the population at large seems to be less affected by economic weakness than one might popularly suppose. People, it turns out, are resilient. An alternative way of looking at it is 'misery loves company' and people do not suffer in adversity as long as their neighbours are suffering as much. 'Keep calm and carry on' is not just a slogan for a T-shirt. The British wartime propagandists who came up with the phrase knew what they were talking about.

Even when the extremes of recession have faded, one consequence of the financial credit crunch is likely to be an ongoing change in consumers' working patterns and a population that tends to be increasingly time poor. This leads to a lifestyle that is not only bad for the environment but also unbalanced in many ways – with problems of insufficient exercise and an unhealthy diet leading to physical health problems. The connection between consumption patterns and health, at this point, is well enough established to need no further discussion, except for one further issue: mental health.

Encouraging financial credit crunches to mitigate environmental credit crunches on health grounds

The typical consumer lifestyle of the 21st century raises questions of mental health, because it not only tends to separate the individual from the environment (perhaps explaining why people are comfortable to go on using resources as if they are unlimited), it can also result in a separation of the individual from family and friends; hence, the frequent discussions of 'work–life balance' in the corporate sector. A field of work known as 'happiness research' focuses on the relationship between happiness and wealth and, although such research often raises more questions than it answers, one of the clearer conclusions is that, beyond a certain level, increasing wealth does not increase happiness.

From an environmental perspective this is an interesting conclusion because it suggests that, once income per head has *exceeded* a certain level, there is headroom for a *reduction* in consumer spending without reducing people's well-being, but with potential environmental benefits. Moreover, because altruism is thought to make at least some people happier, establishing a strong association between the health and well-being of people who are important to us, a healthy environment may well turn out to be quite an efficient social mechanism for a redirection of investment of dollars in the direction of the environment.

Should this conjecture turn out to have any basis in reality, making such an association as well as designing the urban environment to permit a closer connection to the natural environment could have the important side-effect of improving people's mental health. Mental well-being is an unsung but very important dimension of health in a more general sense, and the absence of it can be a significant cost to health services and employers in developed countries.

Do we really want to live the way we live now?

The patterns of work and stress that have been established by modern society are unlikely to be reduced by the financial credit crunch. As consumers react to insecurity by seeking to shore up their positions through harder work (categorized as 'increased productivity' by economists), the financial wealth that is secured has to be offset against a time deficiency, with all that that implies. The patterns of work feed into behaviour patterns – in buying ready meals, labour-saving devices, bundled fast goods and so on – which have negative environmental implications. These products are purchased, not quite irrespective of the environmental consequences, but with a sense that the priorities of personal well-being take precedence over the larger but more abstract notion of environmental well-being.

Prevention versus cure: Spending patterns in the twin credit crunch era

The benefit of prevention (rather than cure) is a clear subtext in this chapter. Although a well cared for environment is not necessarily always part of the answer, it is often present in an indirect manner. In general, it seems to be possible to address a number of potentially very significant health problems at the same time as addressing some significant environmental problems. From an economics perspective, the possibility of killing two birds with one stone or, in effect, making each invested dollar deliver in several directions, is attractive. Conversely, the costs of 'cure' (or when cure is not possible, problem management over the long term) tend to be high whether viewed in financial or social terms.

Preventing health problems could come through preventing the environmental problems we met in the first section of this chapter. Preventing over-consumption, the subject of the second section, could also reduce health-care costs (this point is well illustrated by two recent reports for the UK Treasury that 'produced huge estimates of the value of preventing diet-related ill-health by changing diet' (Lang et al, 2009, p172). (No wonder some governments have been considering the idea of 'fat taxes'.) Finally, preventing adverse working patterns from forming, though perhaps more contentious (or relying on self-discipline rather than an external force), could also prevent health problems from arising.

When the health of a population is not well looked after, this will at some stage turn into a cost for someone (government, tax payer, individual or family members) and also potentially crowds out activities in the health-care system that would, in an ideal world, be directed towards the development of therapies and other health-care technologies of the future. As suggested above, policies that would at one and the same time be good for human health and the environment hold significant potential in terms of better-directing the resources of stretched health-care systems.

So, how does the financial credit crunch bear on the issues of prevention in health care? Health care has traditionally be classified as a 'non-cyclical' aspect of the economy, suggesting that there is some resistance to cutting back in the face of economic adversity. In fact, health-care spending emanating from the private sector is not quite as immune to the economic cycle as is traditionally supposed. While the normal ups and downs of economic activity will produce little impact on the overall spend on health care, a more significant correction (on the scale of the current financial credit crunch) does have an impact.

Personal preventative spending

A financial credit crunch necessitates a reduction in consumer spending and this has the potential to be broad in its scope. The wider definition of health-care spending will be affected by a financial credit crunch.

Spending on medication will respond to a severe economic downturn. Medication covers a wide range of treatments, of course, but within that there is very often a subsection that is not essential. Treatments for common colds, for mild pain and preventative medication (such as multivitamins) – these are all aspects of the consumer basket that can conceivably be forgone. The consumer will willingly cut back on 'non-essential' spending on medication if they feel that they are not absolutely essential to their well-being. This is an example of overconsumption of health care being reined in by economic necessity.

This economizing also applies to health-care related issues outside of the direct area of medication. There is a tendency to cut back on spending on things such as gym membership – or to seek to trade down. The consumer substitutes the expensive gym with the local government sports facility, or even with a vague promise to one's self to take up jogging as an alternative to having a personal trainer. (The road to hell is paved with good intentions.) Real spending on sporting and recreational services (gyms) fell around 10 per cent over the course of 2008 and 2009 in the UK.

A survey of South Korea in the wake of the Asian credit crisis of the late 1990s demonstrates the split effect of health-care spending quite well. Spending on health-care services did drop – by 11.6 per cent. However, spending on drugs (including vitamins and other health-care related products) fell 27.8 per cent. Both in absolute terms and proportionately, South Koreans responded to their more constricted financial circumstances by cutting back more aggressively on their spending on drugs. Spending on visits to hospitals and similar medical services, did not suffer to the same degree.

What is happening here is a reduction in private sector health-care spending. Generally speaking, it is not a cut in spending on 'cures'. If there is something actually wrong with a consumer, then they will pay for treatment. Once a patient has stepped through the door of a hospital, they have effectively become a captive consumer. They will spend on a cure almost without regard to their personal financial circumstances or the prevailing economic condition. They will spend, in short, because the doctor tells them to.

However, in terms of preventative spending, there is a greater inclination to economize. It may be a false economy in the long term, of course, but the financial credit crunch takes priority. For the public sector, there is also an incentive to economize and this might also take the form of economy in terms of prevention.

Governments and preventative spending in the wake of a financial credit crunch

The implications of the financial credit crunch for government spending on health are much the same as for other areas of government spending. In a more credit-constrained environment, with government deficits having increased substantially as a result of the policy response to the economic slowdown, fiscal restraint will require prioritization. This may well reduce government commitments to spending on health. Certainly, it suggests that governments will look for more efficiency in allocating scarce resources to the health-care sector. While the UK's fiscal austerity has specifically excluded 'front-line' health-care spending, it is clear that the increase in spending of the past decade is unlikely to be sustained.

The issue of government spending on health care is subject to the same sort of constraints that limit private sector spending. In a financial credit crunch, governments (as consumers) have less money to spend. Governments may choose to economize on their health spending, in much the same way that the consumer will choose to economize. There is, however, one critical difference between the two forms of spending. The asymmetry of information available to individual patients (who lack knowledge, and are more inclined to spend according to their doctor's direction) makes spending on medical services relatively immune to the vagaries of the economic cycle. That asymmetry does not apply to government. Governments decide health care in the abstract, not face-to-face with medical reality.

The patient walking into a hospital is effectively committing to pay whatever bill he or she is presented with at the end of the treatment. The government, however, does not walk into the hospital. The government decides in advance what it *will* pay (or the formula by which it will determine what it will pay). If a government decides it will not pay for a particular form of treatment, then that treatment is not going to be consumed – regardless of the level of demand that exists for it.

Therefore, any attempt at economy in government health-care spending is likely to result in a reduction in spending on health-care services as well as on drugs. There is one final issue with government spending on health care, however. The government is more than just a cheque book. The government (as the banking sector knows only too well) is also a regulator. As the government seeks to enhance the efficiency of what it spends on health care (whatever is happening to that total sum), the incentive to use regulatory powers to assist in government spending efficiency is likely to increase.

Because the government is seeking efficiency in health-care spending, there is a risk that the government will seek to use its power as a regulator – and ultimately as a regulator of prices. Patent law, fixed prices for medication, regulation of health-care insurance and indeed regulation of medical training all give the government a variety of levers that can be pulled, to exert influence on the efficiency or cost of the health-care sector. Drug pricing is perhaps the clearest example of how this might be significant. If the government chooses to use its regulatory authority to control drug prices, it will change the incentives for pharmaceutical companies to invest in drug development. This is not to say that the regulation of prices is unwarranted – it may well be justified. Whether justified or not, however, it will change the incentives for drugs companies.

This plays into the economics of drug development – by either reducing the potential returns a drug developer may earn or alternatively increasing the regulatory uncertainty associated with the potential return that a drug developer may earn. Reducing return, or increasing the risks surrounding those returns, will reduce the incentive to develop drugs. These effects shape the final relationship between the financial credit crunch and the economics of health – research and development (R&D) spending in the fields of medicine and pharmacology.

The government as a regulator beyond pharmaceuticals

The financial credit crunch therefore imposes costs on both the public and the private sector. However, the government has another preventative role that it can exercise in health care, beyond the provision of health care itself. The government has a role through its position as regulator of pollution.

The Report of the President's Cancer Panel (April 2010) in the US comments as follows on the relationship between cancer and the environment: 'With the growing body of evidence linking environmental exposures to cancer, the public is becoming increasingly aware of the unacceptable burden of cancer resulting from environmental and occupational exposures that could have been prevented by appropriate national action.'

In Europe, the legislation known as REACH (Registration, Evaluation, Authorisation and Restriction of Chemical Substances) 'aims to improve the protection of human health and the environment through the better and earlier identification of chemical substances'. What comes through clearly from both sides of the Atlantic is the lack of knowledge in relation to man-made chemical substances that are found in the environment in quite large amounts. This implies a need for a systematic process of identification, testing and replacement of hazardous substances over a period of time.

So what does the financial credit crunch have to do with this aspect of health and the environment? Directly, regulation is an appealing policy option, as it is relatively costless in itself. However, the need for identifying and testing hazardous substances over time is not likely to be cheap. This is, unfortunately, exactly the sort of programme that could come under pressure in the race to cut spending, post crunch. The US report cited above refers to inequity across the wealth divide, in terms of which segments of the population tend to be exposed to dangerous chemicals. The implication of this may be that, the lower the GDP per capita, the less money there is to go round on scientific research and other investment designed to protect human health from dangerous substances in the environment. Looking back to the earlier section on environmental degradation, it is clear to see that any such reduction in GDP is likely to set in motion effects that could potentially affect the health of future generations for years to come. As with many health problems, prevention would clearly be better than cure.

Prevention from research and development

The implications of the financial credit crunch on research and development were summed up by the poet Whittier: 'of all sad words of tongue and pen, the saddest surely are these: it might have been'. This is somewhat more succinctly phrased by the economics profession as 'opportunity cost'. The problem with the financial credit crunch interacting with the economics of health-care R&D is not that existing health is negatively affected, but that *future* possibilities are not discovered.

The cost of capital (again) and the economics of health

The financial credit crunch is largely about the cost of capital. As we saw in the chapter on infrastructure (Chapter 4), the rise in the cost of capital that is the consequence of the financial credit crunch will reduce the ability of companies to spend *now* in expectation of a return in the *future*. Pharmaceutical research and development is much the same as infrastructure spending. Rather than building a physical infrastructure of roads, bridges or factories, we are building a mental infrastructure of a stock of knowledge. The R&D necessary to increase our understanding costs money up front, and the return on that knowledge may only be harvested over a number of years. This, of course, is why pharmaceutical companies are so wedded to their patents (and why the expiry of a patent is such a significant event for the sector).

There is an added complication with medical R&D: the outcome is less certain. With infrastructure spending, the return has an element of uncertainty, to be sure, but the absolute usefulness of the infrastructure spend is rarely called into question. One might say that infrastructure created is more or less useful, and will generate a higher or lower return over time. It is rare that infrastructure will generate no return in the future.[5] With pharmaceutical R&D, the commitment to spending on research now is more speculative as to the likely return. Certainly, a breakthrough discovery could have profound (and profitable) consequences in the future. Equally, years could be invested in pursuing will-o'-the-wisps of medical insight, with no commercial conclusion.

What medical (pharmaceutical) research therefore represents is an illiquid investment, which pays returns over a long period of time and whose returns carry considerable risk and uncertainty. If we juxtapose this set of circumstances with the financial credit crunch, where liquidity is desired (and illiquidity shunned) and where risk attracts a higher cost, it can be little surprise that R&D spending by pharmaceutical companies has been constrained.

In fact, the period after the financial credit crunch (to 2010) marked the first time in 40 years of recorded history that pharmaceutical R&D spending has *failed to increase* in nominal terms. In real terms, of course, spending has been declining.

As well as the issue of commercial health, we can perhaps add the overlay of the charitable sector. Medical charities are important in financing (and even conducting) research at an academic level. The role of charities in financing cancer research is an obvious area where this occurs. However, charitable funding of research is affected by the financial credit crunch, just as much as any other aspect of the economy. First, the ability (or, perhaps, the willingness) of the population at large to support charities is constrained in more constrained times. Charitable donations in the UK fell 11 per cent in the 2008/2009 fiscal year, reflecting the consequences of the financial credit crunch. Furthermore, medical charities are likely to have to compete more vigorously with other charitable endeavours in terms of raising funds. For instance, if government support for education is reduced as a consequence of fiscal constraints in the post financial credit crunch era, then educational charities may seek to increase donations from their supporters. This puts them into direct competition with other charities, including medical.

As if this were not enough, charities are also affected by the financial markets. Charitable endowments are affected by the decline in asset values. Furthermore, in an environment where investment returns are lower (and expected to be structurally lower as a direct consequence of the financial credit crunch), charities will have to adjust to a world where their investment income is more constrained. Charities tend to be biased to lower-risk investments. This may mean that the capital losses of the charitable sector are less than other sectors. It also means, however, that the income earned from those investments will be lower (low-risk investments such as good-quality government bonds are generating lower yields as a consequence of government policies, liquidity preference and other distortions in financial markets).

So what does this mean for health? We are back to the 'saddest words of tongue and pen' again. There is no change to the current status quo for health. The existing stock of knowledge is unlikely to be forgotten – short of some global catastrophe. However, the potential to improve health in the future is (on balance) likely to suffer. A cure for Alzheimer's has to be assumed to be less likely, if there is less money being spent in pursuit of a cure for Alzheimer's.

(Of course, there is still the possibility of chance discovery.) At the same time, the probability of medical advance is reduced in aggregate by the reduction of funding availability, from whatever source. What might have been discovered remains undiscovered – undiscovered for longer or perhaps forever.

Living with the 'ménage à trois' – finding the right mix of environmental and health-care investment

As this chapter has suggested, health is a complex business. The environment in which we live (produced by the economy that supports how we live) has a bearing on how healthy our lives are. The way we live now has shaped our consumption patterns (towards overconsumption, within the OECD), which feeds into both the environment and health. The way we live now also shapes our working patterns, with implications for health and the environment.

The challenge is that the financial credit crunch is changing the way we live now, but it is not necessarily doing so in a way that is beneficial to health or to the environment. More concerted change in a different direction may be required.

It is worth considering that small changes in the scheme of things can potentially make significant differences. As an example, more than one city has recently introduced a rental bikes scheme. They are not necessarily the ultimate in 'green' transport – on the whole, the life-time environmental footprint of bike renters will be higher than that of the urban walker because of the resources required to manufacture the bike and also those needed to maintain the infrastructure that makes the bikes available. Nevertheless, they rank highly for the multiple benefits they bring – including health. For one thing, in significant numbers, the users of Boris's Bikes or their equivalent might potentially improve physical conditions for others – by not belching particulates and other pollutants into the atmosphere from a car exhaust pipe, or by not adding to the congestion in bus or tube. For another, using human energy to make short urban journeys is likely to bring significant health benefits to the individual user, by introducing regular exercise into an otherwise sedentary way of life.

The message of this chapter on health is that to focus purely on health services provision, is likely to result in an overuse of the resources of the health-care sector, as well as a misallocation of scarce (health-care sector) resources towards preventable ills rather than the ills that truly need fixing. More than one catalyst may be needed to bring about changes in the way health policy is planned and health services provided but, in light of the depth of spending cuts developing in some countries, the financial credit crunch may turn out to be quite a heavy-hitting catalyst in the context of changes to lifestyle, that is helpful to human health and the environment at the same time.

The fate of investment needed to deal with the potential health effects of environmental degradation is another matter – sadly, research tends to be an easy victim in the context of cost cutting. Does this matter? It depends. A glance across to the chapter on fast goods (Chapter 7) suggests that if the health risks of environmental degradation are widespread enough, then economics can be brought to bear to mitigate them. However, the eventually successful R&D in the context of our foamy drinking water example focused on replacing a substance without which a very large consumer product could have been under threat. The threat that moved companies was their commercial well-being rather than an unknown health effect.

We are not saying this is a bad thing: the right solution (removal of the offending substance) was reached. The most rapid pathway was certainly through commercial R&D – finding out (first) whether or not human health would be affected would have taken a very long time and, by the time the answer was reached (if ever), a lot more of the non-biodegradable substance governments wanted to see removed would have been irreversibly out in the environment. This can be seen, in hindsight, as a case of efficient use of R&D dollars from a health perspective. However, this does not mean everything can be left to the private sector, as someone needs to make sure incentives are properly aligned.

As a counter-example, the 2008–2009 Annual Report of the US President's Cancer Panel (published in April 2010) observed:

> *Research on environmental causes of cancer has been limited by low priority and inadequate funding. As a result the cadre of environmental oncologists is relatively small, and both the consequences of cumulative lifetime exposure to known carcinogens and the interaction of specific environmental contaminants remain largely unstudied. There is a lack of emphasis on environmental research as a route to primary cancer prevention.* (Lefall and Kripke, 2010)

The pharmaceutical industry has tended to direct research dollars towards genes and molecules rather than the environment. The gains from identifying environmental causes of cancer would be more likely to be reaped by a range of stakeholders – there is no patent to secure and no stream of identifiable royalties for the commercial sector. The question is, then, who should pay. The answer, in a financial credit crunch, is likely to be 'no one'.

This chapter draws three broad conclusions. First, the environmental credit crunch is negative for human health (therefore something that should be integral to any health policy). Second, financial crunches can affect human health both positively and negatively, depending on specific behaviour and resource allocation. Third, financial crunches will affect health-care spending: this need not be negative if unnecessary spending is reduced, but the impact on investment (in health-related R&D) does not look so favourable. Squaring the circle, the most important answer to drop out of the discussion is that integrated thinking in respect of the relationship between the environment, economics and human health, and the way they interact, is a must if human health is to be best looked after in the 21st century. Prevention and pre-emption are decidedly better than cure when it comes to health care and the environment, whether singly or in combination.

Notes

1 The problems experienced by householders in Ireland at the end of 2010, when the 'big thaw' revealed weaknesses in the water infrastructure, said to be a consequence of underinvestment over a long period of time, is a good case in point. As health professionals pointed out at the time, rising health risks are an immediate consequence of insufficient access to clean water.
2 The source is the President's Cancer Panel, 'Reducing environmental cancer risk' (National Cancer Institute, 10 April 2010).
3 BBC Health – Understanding Asthma. See www.bbc.co.uk/health/physical_health/conditions/in_depth/asthma/aboutasthma_index.shtml.
4 Emile Durkheim observed that suicides tend to increase during economic changes that disturb 'the social fabric' as long ago as 1897. Economists would contend that the financial crisis is an event that is clearly disturbing the social fabric today.
5 An example outside of health care is the canals of the 18th and early 19th centuries. These forms of infrastructure continued to generate some positive return for many years after they were superseded by railways as a means of transporting goods. Although the returns may not have matched the initial expectations, returns were there to be made once the infrastructure was in place.

References

Lang, T., Barling, D. and Caraher, M. (2009) *Food Policy: Integrating Health, Environment and Society*, Oxford University Press, Oxford

Lefall, L. and Kripke, L. (2010) *Reducing Environmental Cancer Risk*, US National Cancer Institute, Bethesda, MD

Pensions Commission (2005) *A New Pensions Settlement for the Twenty-First Century: The Second Report of the Pensions Commission*, The Stationery Office, London

Smith, D. (2008) *The State of the World Atlas*, 8th edition, Earthscan, London

CHAPTER 9

Education, Work and Leisure in the Face of Two Credit Crunches

You dropped a hundred and fifty grand on a fucking education you could have got for a dollar fifty in late charges at a public library. (Will Hunting in *Good Will Hunting*)

Apart from a certain amount of time for sleeping and eating, people are either in education, in work or at leisure, or some combination thereof. Even eating has its leisure aspect, given that (in the Organisation for Economic Co-operation and Development (OECD) countries at least) eating has long since surpassed the task of acquiring the calories necessary for survival. Whether to go with Jamie Oliver or Turkey Twizzlers is a lifestyle choice, not an essential facet of human existence.

An event that is as wide-ranging as a financial credit crunch must touch some or all of these occupations. Indeed, all of the preceding chapters of this book could be said to be addressing issues of education, work and leisure – albeit in a somewhat circuitous fashion. It seems appropriate that the final chapter of this work to look at sectors of the economy should examine the specific impact of the financial credit crunch in these three areas.

At the same time, given that human activity is always engaged in one of these three occupations, the consequences of education, work and leisure for the environmental credit crunch must be considered at the very heart of the current challenge. If people change the way that they work, it has a bearing on the environment. Think of the environmental impact of the urbanization of the Industrial Revolution in the UK – or the predicted urbanization of emerging markets. Moving from cottage industries to factories (or rural work to industrial work) changed demands for transport, food, water and construction. It had profound impacts on health. It also changed the demands for education (industrial workers needed to follow written instructions) and demands for leisure.

So, how does the financial credit crunch change the way in which society approaches education, work and leisure? And what does that mean for the environmental credit crunch we now face?

The financial credit crunch and education

Education encompasses a good deal. There are schools and universities of course. That much is obvious. But education pre-dates school – indeed, pre-school education is a highly profitable business. Education continues for adults – not only through formal adult education schemes, but through training in the workplace and indirectly through access to books (possibly including this tome) and other media.

The economics of education in the wake of the financial credit crunch

The changes that the financial credit crunch will bring about to education might raise a sense of déjà vu from the discerning reader. There is good reason for this. Much of the discussion about education in the wake of the financial credit crunch parallels the earlier chapter on infrastructure spending. Infrastructure is, of course, investment in physical things. Education is simply investment in the quality of labour (labour and physical capital being the two most fundamental components of an economist's world).

So what does the financial credit crunch entail for the economics of education? Essentially, this comes down to state funding, personal funding and corporate funding. All are likely to be adversely affected.

State funding of education is certainly more likely to be constrained in the wake of the financial credit crunch. This is true of all forms of state spending, of course. A slower-trend rate of growth combined with the burden of increased government debt will restrict the ability of a government to spend on anything. How the burden of budget restraint is allocated is a matter for politicians. With education accounting for around 13 per cent of the UK's general government budget and some 16 per cent of the US general government budget, it seems plausible that some burden of the cuts will fall on the sector.

Reduced state spending on education either means that there is less spent on education per person, or that the number of people who have access to state-funded education declines. The former has been evident in the US, where spending per pupil in higher education is now at a quarter century low (mainly as a result of state government budget constraints). The current controversy over undergraduate university fees in the UK is essentially a facet of the latter argument – the number of people who have access to *fully state-funded* higher education is likely to dwindle.

State funding of education extends beyond schools and undergraduate studies, and into postgraduate research. Here, too, there is likely to be a degree of economy. State-sponsored research can be an important part of an economy's ability to innovate (the computer owes its roots to state-sponsored research as, of course, does the internet). The interrelationship of research, innovation and the economy is not fixed. Research led to the innovation of anaesthesia in surgery but, while undoubtedly a humanitarian benefit, there is no evidence that this innovation directly led to economic growth.

The financial credit crunch is thus likely to reduce state funding of education overall. Unless this is substituted with other sources of funds, this will either reduce the quality of education output ('dumbing down' potentially) or reduce the breadth of education (academic elitism).

Household funding of education in the wake of the financial credit crunch

Clearly, the state is not the only source of funding for education. Education is also paid for by individuals. In the US this is obvious – college tuition may sometimes be paid out of grants and bursaries, but for the most part it is the responsibility of the individual to finance their own education. This can happen with cash up front or with a deferred payment scheme (otherwise known as a loan). In the UK and Europe, education is also paid for by individuals, to a greater or lesser extent. In some cases (as with the UK), this is through loans that are to be repaid when education is completed.

In other instances (technically, in all cases), time spent in education incurs the cost of earnings forgone. Forgone earnings should not be disregarded. If there is no tradition of higher education (in particular) in a family, then the potential to earn money in the short term could well carry a greater weight than the abstract benefits of continuing with education.

We must not confine ourselves to too narrow a view of household funding of education. Households will also pay for education by straying into grey areas with other areas of consumer spending. Educational toys may be purchased for children. Computers may be bought whose first purpose is education. Even visits to museums, historic properties and galleries could be considered educational.

The two forms of household education spending – what might be termed 'direct' and 'indirect' education spending – will be affected by the financial credit crunch. However, the economics is likely to be slightly different in each case.

For direct household spending on education, there are two questions that arise. First, if (higher) education is to be credit-financed, is that credit still available? Although state-guaranteed loans are available, they generally cover only tuition fees. To the extent that one needs to borrow to support oneself while studying, this form of credit may be less readily available in the wake of the financial credit crunch. A loan to a student is a higher-risk proposition for the private sector to undertake – essentially the lender is making a bet on the future earnings potential of the borrower. If risk aversion has risen in the wake of the financial credit crunch, then this is a problem.

Second, is the potential student prepared for the time spent on and expense of education, if the certainty of employment at the conclusion of that education has declined? In 2010, the media in the UK were full of reports about the level of graduate unemployment (reported as being 14 per cent of recent graduates in December 2009. US college graduates meanwhile face the highest unemployment rate in four decades). Of course, college graduates have better employment prospects than their lower-skilled peers, but from the vantage point of someone contemplating three years of expense with little immediate income, this may be of limited comfort.

In assessing the direct spending on education, therefore, the question is whether the financial credit crunch restricts the ability to purchase education (just as it restricts the ability to purchase any other good) and whether it creates a disincentive to continue in education.

For indirect spending on education, the issue is more directly related to spending behaviour. If consumers are more financially constrained, or their income is growing less, they may reduce their spending on education as on any other item. Buying educational toys may seem less of a priority to a credit-constrained household. Someone who is out of work is less likely to use their leisure time to visit museums and historic monuments, given that this will incur a cost in terms of travel and possibly admission.

At the same time, we have to recognize that indirect spending on education is just one element of the consumer's shopping basket. If the consumer prioritizes education, it may be at the expense of other items. Certainly, anecdotal evidence suggests that indirect spending on education for children has not notably diminished (for example, the children's educational toy manufacturer LeapFrog has reported rising sales throughout 2010).

Household spending on education seems likely to resolve in a discrete pattern. If it can be afforded, it will be purchased. If it cannot be afforded, the financial credit crunch reduces the likelihood of it being purchased. What this means is that, as far as household spending is concerned, the broadest definition of education spending is likely to become more income-dependent. Middle and higher-income groups will be able to afford the indirect spending on education. They will be more readily able to access non-state-supported forms of credit that fund direct spending on education. Lower-income groups are less likely to have the flexibility to be able to continue (or embark upon) indirect spending on education. They are also potentially less able to embark on direct spending on education, either because of the loss of potential income or because the uncertainties surrounding future employment make the risks associated with the spending too high.

Summing up, the economics perspective suggests that the financial credit crunch is likely to reduce state funding of education, and to render private spending on education income-dependent.

Environmental perspective on education

The state of affairs described above depicts a world in which key decisions about educational provision will be reshaped by the financial crunch. Education is, in an important sense, no different to the environment: both can be described as strategic issues for government and consumer, the health of both being fundamental to the stability of the economy and to the well-being of the individuals within it. Government policy decisions and consumer decisions will affect both dimensions – economic performance and societal well-being – and in the context of the financial credit crunch, two important tensions between them emerge. In crunches, what can be counted (i.e. quantified in numbers) will tend to count (i.e. have greater weight in policy); and government policy-makers and consumers alike will tend to focus on what can be afforded in the short run.

The value to the economy of investments made in education (as indeed in the protection of the environment) can be hard to measure. It can thus be hard to defend educational investment when the government decides belt-tightening is required. As an example, the gradual disappearance of school sports fields in recent decades can be said to have had environmental and social costs – perhaps a fall in biodiversity where the field was, alongside a rise in obesity and a fall in mental well-being for the children concerned.[1] However, none of this can be easily proved. Each individual sports field involves small numbers, which compounds the difficulty of arguing for environmental and social value on a case-by-case basis. As this suggests, financial pressures will tend to have similar effects on the long-term stewardship of both educational and environmental assets, potentially putting both at risk by focusing everyone on the short-run benefits or costs.

A key difference between education and the environment is that, whereas more economic activity (other things being equal) tends to be bad for the environment, strong economic growth will tend, all other things being equal, to be good for the volume of educational provision (because funds to support it are more likely to be available). However, both the negative environmental consequences of economic growth and the increase in potential educational resources that stronger economic growth can bring will tend to be short-run and random in their impact. What really matters, anyway, is strategic planning. The effects of expansion can be wasteful – obviously in the case of economic growth, but also in the case of education. Waste occurs if activity exceeds what is required and that misallocation of resources (to put it into the educated cadence of economics) will tend to have a negative environmental consequence.

In the UK, a government policy-driven expansion in university places in the late 20th and early 21st centuries took place against the background of benign economic conditions, so that the financial stresses and strains of providing the added capacity were not felt as much as they might have been in normal conditions. Moreover, as people felt wealthier in the years of expansion, private sector schooling received a boost from individual decisions to opt out of publicly provided education. As the financial credit crunch took hold, the prospect of paying more for degrees brought students out onto the streets in protest and the competition for places in public sector schools (not to be confused with 'public schools', which is the name for private schools in the UK) skyrocketed in some regions as parents were forced by straitened financial circumstances to return to the state education sector, with its limited capacity to absorb new (unexpected) students.

With 20-20 hindsight, these swings in behaviour on the part of the government and the consumer could have been reduced by applying a stricter economic discipline on the way up as well as on the way down. In a consumer boom, overconsumption often leads to waste, and thus it is likely that the rapid UK expansion in university places has led to a waste of resources. This is not intended to suggest that more spending on education is a bad thing. How things are done tends to be as important as what is done, and the fact that social mobility in the UK has fallen even as spending on education rose suggests a misallocation of resources somewhere in the system.

This brings us to a potentially controversial point about the volumes of educational provision. Both of the writers of this book come from generations in which the power lay with the relevant educational institutions to select their students on the basis of their assessed performance: A-grade school exam results and first class degrees alike were relatively rare. So-called 'grade inflation' can be seen as one consequence of a shift in the balance of power brought about by the modern tendency to consider 'the price of everything and the value of nothing', in conjunction with a government policy to put more than half the population through university. These trends have arguably turned education into a consumer good – if not in all senses of the word at least in the sense that the ability to afford it financially has become more important than it used to be.

Following the financial credit crunch, the consumers who will have to spend more on their university education in the future may find themselves without the expected quid pro quo of that investment – a well-paying job. The quote from *Good Will Hunting* at the head of this chapter bears repeating: 'You dropped a hundred and fifty grand on a fucking education you could have got for a dollar fifty in late charges at a public library.' People like Alan Sugar and Richard Branson (and come to that literary geniuses of the Elizabethan Age such as Shakespeare and Jonson) are proof (living and historical) that a highly productive education can take place outside a formal institution. Moreover, the environmental perspective would suspect that not relying so much on formal institutional provision might entail a lower carbon footprint somewhere along the line (depending on what activity the years not spent in an educational institution displace).

The aim of a well-run education system in democracies is generally said to include the attainment of reasonable levels of numeracy and literacy (the raw material for work and leisure) in the population, together with the provision of advanced education in such a way as to make it accessible to those with raw talent. The fall in social mobility that some OECD societies have experienced suggests that access to the system on the basis of ability is not always being delivered even though provision has expanded. If this is correct, the real importance of this loss of social mobility is that people who might have become mathematicians, biologists, doctors, engineers, physicists or environmental scientists (which entail high barriers to entry in the form of a high level of knowledge as well as the raw talent to use it) may have been lost.

The first we will know of this loss of knowledge will be in decades to come when the number of UK patent applications drops down the global league table. If this happens it is worth emphasizing that this will not be because of the post-crunch retrenchment underway. Instead, the damage to the knowledge stock may well arise from the rationing of education (rationing through the mechanism of pricing), which was already becoming more important before the financial credit crunch hit the economy. This suggests an unfortunate parallel for the environmental crunch – in which an undisciplined exploitation of environmental resources has been applied over decades, suggesting a painful (and potentially inequitable) reversal may be required at some point.

Rather harder to capture than the interaction between education and politics (and in parallel politics and the environment) is the way in which the environment and education interact with each other. However, perhaps the key point for this chapter and indeed this book is that what we consumers understand about the environment has a strong influence on what we consumers do to the environment in the long run. In this broad sense, the state of the education in any given country has some relevance to the environment, for education shapes what we know and what we do with knowledge, and this in turn will help determine the future evolution of the impact of the relevant economy upon the environment.

If the value systems inculcated through what we learn in our formative years fail to embed some of the concepts discussed in earlier chapters (such as entropy), the poorer-quality education that results will risk having (albeit indirectly) a very negative impact on the environment. Such an education system still allows an improvement in economic productivity (assuming the workforce is literate and numerate), but does so without fostering the necessary environmental checks and balances.

With a shift in attitudes, almost the same educational content in terms of the sciences and the arts could be deployed in a much less damaging way for the environment, by changing the way things are made or used as well as what we spend money on as consumers. Looking back through previous chapters, the development of the know-how and technology needed to procure, distribute and use energy is just one example; for the way energy is procured and used is driven by the formerly held assumption that cheap energy would always be a permanent feature of the landscape. In the modern age, nascent efforts are underway to develop clean energy and students are learning about the current state of play in knowledge of the environment. All of this suggests things could change significantly in coming decades, as generations more enlightened about the importance of the environment enter the economy and apply the disciplines of mathematics, physics and engineering in a way that takes the impact of economic activity on the environment into account and (as a minimum) tries to avoid techniques or processes that cause environmental depletion.

In fact, education is fundamental, not simply for the avoidance of environmental damage, but for the delivery of environmental solutions of the future. In previous chapters we have mentioned from time-to-time the importance of the ability to develop new technology in order to develop economically. The urgency, for the environment, is that new clean technologies need to replace the old ones, and this means that a sound education in the sciences and the humanities, along with a system open to the most talented on the basis of ability (rather than the means to afford it), has never been more important. As Peter Day (BBC Radio 4 website, 25 November 2010) has written in the context of an interview with John Kay, the one thing that seems to be able to keep growing without discernible limits is human ingenuity. In fact, paraphrasing Kay's argument, the only limit to human ingenuity is … human ingenuity. This is why, to go back to the Kenneth Boulding quote that led the Preface to this book, economists are not mad, even if they assume infinite potential in a finite (physical resource) world. Human ingenuity allows us to do more with the same. It can also (as the environmental credit crunch bites) potentially allow us to do the same with less, or maybe even more with less. Human ingenuity and finite resources can still create infinite possibilities.

Looking back through history, civilizations have tended to flourish at times when education has been at its most highly prized – for instance, the Elizabethan expansion in trade was facilitated by a significant expansion of school and university provision under Elizabeth I. We cannot say which is the chicken and which is the egg of course but, for our purposes, that does not matter – what *does* matter above all for sustainable growth is that economic retrenchment must not diminish the quality of educational provision and also needs to be done in such a way as to improve equality of access on the basis of ability. After all, our educational systems create the pool of academics that push forward science, the arts and the humanities, as well as the pool of people that understand how to make use of them in practical terms. The question is what will happen to this important dimension during the financial credit crunch.

One consequence of financial downturns and credit crunches is that many social expectations that have become implicit contracts (such as retirement at 65 in some countries) have to be 'defaulted' on, because they have become unaffordable,[2] so that the ones that are really needed can be kept. Thus, as we write, the debates in the UK on pension fund provision and retirement ages, and the structure of local authorities, health services, education provision and environmental policy are progressing in parallel and forcing difficult decisions on organizations and consumers. Under pressure, it is unlikely all the 'right' choices will be made everywhere (that is potentially a book in its own right).

Education and the environment alike are vulnerable to potentially perverse decisions. Currently, students and educators in the UK are pushing back vociferously on the move to increase fees. They are facing an uphill struggle – their arguments are (at the time of writing) not gaining traction against the counter arguments (that only those who can afford it will be required to repay the larger loans needed to pay fees).

The difficulty with gaining traction is quickly discovered if we contrast this with what is happening to the proposed change to the regulation known as the Carbon Reduction Commitment (CRC), which was quickly recognized as needing second thoughts. Under the relevant regulation (which came into force in April 2010), organizations using more than 6000MWh per year will be required to record and report their carbon dioxide emissions from July 2011 onwards and, starting from 2012, will have to buy carbon dioxide emissions allowances to cover the previous year's emissions. The original design of the scheme mimicked the European Emissions Trading Scheme, in which entities successfully cutting their emissions would be able to monetize the saving by selling permits back to the market, thereby retrospectively covering the costs of investments made to reduce their carbon footprint. In the UK spending review of October 2010, an important element of the CRC changed: 'Revenue from the sale of CRC allowances, totalling £1 billion a year by 2014/15, will be used to support the public finances, including spending on the environment, rather than recycled to participants.'[3] Leaving aside the medium-term misalignment of incentives this change could create, the immediate effect was an additional cost for companies right in the middle of the financial credit crunch, especially for those that had 'done the right thing' and had already spent money on investments intended to mitigate their carbon emissions. As a consequence of the relatively short-run impact on finances for companies, push-back was immediate and a rethink is underway. In financial credit crunches, money talks.

Thus, speaking in general terms, financial credit crunches will tend to put both environment and education policies at risk if investment in them requires any sort of funding in the short run for them to be sustainable in the longer term. While recognizing that careful financial stewardship is just as important in education or the environment as anywhere else, the timing problem is extremely hard to resolve, but not coming to a solution may entail significant risks. After all, the importance of a sound education system is that it will render people better-equipped to deal with big changes – whether they are the medium-term changes in the economy that will emerge in the aftermath of the financial credit crunch or the longer-run structural changes that will arise as a consequence of the environmental credit crunch.

The economics of employment in the wake of the financial credit crunch

From education, we naturally progress to employment. Moving from education to employment is the normal progression in the course of one's life or, at least, that is currently the expectation in developed countries. The interrelationship between education and employment, and the consequence of the financial credit crunch, are intimately bound up. For an economist, this is potentially one of the most interesting, exciting and concerning aspects of the financial credit crunch today.

As education represents an investment in the quality of human capital, it is closely associated with the skills of the labour force. Skills are something which may make someone more or less employable. Hence, education and employment connect.

In a modern industrialized economy today, skills have become ever more important to employment. Globalization has meant that jobs relating to traded goods sectors (i.e. those parts of the economy that can be imported or exported) are increasingly mobile. The 'outsourcing' so beloved of populist politicians is relevant here. However, it is unwise to suggest that outsourcing is an overwhelming trend in any economy. Large swathes of an economy (e.g. the domestic service sector) are not tradable, or tradable only with difficulty.

The problem comes when there is a wide range of skill levels in a society. An economy that has very high-skilled and very low-skilled workers becomes, effectively, schizophrenic. It is trying to be an advanced economy and an emerging market at the same time. Such a characterization could be applied to the US, for example. The US has a large number of highly educated workers produced by its education system. At the same time, it has a high level of illiterate and innumerate workers. (Functional illiteracy was pegged at 27 per cent and innumeracy at 39 per cent of the population in 2005, according to the National Assessment of Education Progress. A functional illiterate would be incapable of reading and comprehending a newspaper article, for instance.)

During a period of economic prosperity, employment can generally be found for many. High-skilled workers have employment because they are producing valuable output in an economy that has a comparative advantage in high-skilled output. Low-skilled workers find employment because, effectively, the economy is inefficient and employs more (cheap) low-skilled workers than employers actually need; more people serve hamburgers, operate 'hand car-wash' services and otherwise engage in low-skilled occupations.

As soon as a recession hits, particularly one that is associated with a financial credit crunch, there is an increase in unemployment among the lower-skilled workers. This is effectively a correction for the abnormal employment occurring in the economic expansion. However, a financial credit crunch adds its own peculiar dimension to the process. Because a financial credit crunch limits the speed of a recovery, the slower rate of economic growth is unlikely to reduce unemployment (particularly low-skilled unemployment) very rapidly.

The problem that emerges here is that the longer one is unemployed, the less likely one is *ever* to find employment again (the 'hysteresis effect'). Skill levels rust away through lack of use. The unemployed worker sends out the wrong signals to potential employers (along the lines of 'no one else has thought me worth employing, so why should you bother to give me a job interview?'). Of course, the unemployed worker is likely to become dispirited and make less effort to find employment.

One of the features of the financial credit crunch, in the UK and in Europe but most particularly in the US, has been an increase in the numbers of long-term unemployed workers (long term normally means being out of work for more than six months). The long-term unemployed are disproportionately low skilled (most of the long-term unemployed in the US failed to graduate from high school).

So, the financial credit crunch increases unemployment and increases the structural or permanent level of unemployment within that. This creates shifts in income which changes spending patterns. Economy (in the sense of being careful with money) will be a trait of consumers who have either experienced unemployment or feel threatened by its proximity. We have already seen how this trait will influence consumer spending across a wide range of products, with environmental consequences.

The financial credit crunch has more consequences than unemployment alone. Those with employment are likely to experience changing working patterns as a consequence of economic developments. There are two traits that come out of the financial credit crunch – at first glance contradictory but in fact arising from a common cause. For those in employment there is likely to be an increase in part-time working, but also an increase in working hours.

How can two such developments co-exist? The roots of such trends come from two areas. From the labour demand perspective, employers are likely to be concerned about committing to full-time employment at a time of economic uncertainty. Full-time employment entails a longer-term commitment to increasing costs, and full-time employees are generally better protected by labour legislation. Therefore, it makes sense for an employer to first seek to get their employees to work longer hours. There may be a cost associated with this (if the employee is paid on an hourly basis), but even that is not certain (salaried employees do not necessarily get increased compensation for longer hours). If increasing hours worked is not viable, then the next alternative is to use part-time labour (what used to be known as 'casual' labour). There is normally greater flexibility around the employment of part-time labour and it is very often less costly.

So much for the demand side; from the perspective of labour supply, there are again incentives that produce this pattern of work. Existing employees may feel incentivized to offer longer hours from fear of job security. 'I had better work overtime as I might need the money' or 'I had better work longer so as to avoid the threat of losing my job'. At the same time, fear of unemployment may also incentivize people to work part time. This latter trait comes about because (generally speaking) people live in families. Of course, more people do live in single person households today than has been the case in the past, but two-adult households are still widespread.

Where there is a two-adult household that has *chosen* to have only one adult in employment, a financial credit crunch may well induce a change in employment behaviour. If a family is fearful of future unemployment, has actually experienced unemployment or simply is experiencing difficulty in maintaining its lifestyle because of restricted credit or reduced income, there is an incentive to increase employment participation. What happens is that the second adult in the household will seek work.

In most OECD societies, this generally means that female participation will increase (as there is still a bias to women in a two-adult household having primary responsibility for children and being voluntarily out of paid employment). Where issues such as childcare remain a concern, there is therefore a strong incentive for the increased participation in the workforce to be on a part-time basis. What is happening here is that the household is going from being a one-income household to being a one-and-a-half income household, in response to the constraints or fears that arise from the financial credit crunch. We saw in the opening chapter that this sort of change in working patterns will have further implications for consumer spending on things such as food.

So, the financial credit crunch produces three clear labour market trends:

1 There is an increase in unemployment, particularly low-skilled unemployment, which is likely to be permanent.
2 There is an increase in part-time working.
3 There is an increase in hours worked.

Environmental perspective on employment in the wake of the financial credit crunch

Consumers are also employees, so what happens in workplaces will affect them in two dimensions – spending power in their personal lives (as jobs are cut or benefits reduced); and the quality and quantity of output delivered by their workplace in the presence of cost cuts. Put together, enforced frugality in the home and the workplace could be seen as a potential catalyst for a more environmentally friendly approach to the product and service outputs work delivers, as well as to the way work itself is done. (But, not necessarily.) For those fortunate enough to keep their jobs post-crunch, cost controls in the workplace may have the potential to stimulate the development of 'human capital' in the broadest sense of the word as people use their ingenuity to deal with cost constraints. This could also bring about environmentally helpful changes in the way goods and services, and the resources required to deliver them, are procured and used. (But, not necessarily.)

'Not necessarily' because here, as in previous chapters, we do not find any certainty that post-financial credit crunch frugality will also mean environmental frugality, even in the face of the environmental credit crunch. What is needed for the response to the two crunches to evolve in a coherent fashion with respect to the environment is a shift towards a stronger association between economic or financial frugality and environmental stewardship. Whether this happens will depend as much on (technical and other) training in the workplace as on credit crunch-driven shifts in thinking. For, as should be clear by this stage in this book, unless environmental thinking is embedded in the decision-making process relating to cost cuts, the environmental impact of the cost cuts will be random – sometimes good, but sometimes bad.

When it comes to the direct effects of working patterns, work (defined as what we do in paid employment) triggers activity that can potentially increase the household's environmental footprint. The most obvious environmental impact arises from the transport requirements associated with job location. Although many jobs have to be done 'on location' (in an office, factory or shop) changes in technology do allow some service sector employees to work 'anywhere' (which generally means 'anywhere with a suitable internet connection'). This is almost an anti-industrial revolution, recreating cottage industries out of parts of the service sector. As a topical example, parts of this book were written on planes and in trains, and the authors frequently (constructively) debated content while located at opposite ends of the planet. When such flexible working patterns become widely accepted for service industries, allowing employees to work with greater autonomy, it permits less energy intensive ways of working. If a company only employs home workers, less total floor space (home plus office) is needed, potentially translating into a lower consumption of heat and light. The energy used in travelling to work is considerably less as all that is required is to wander downstairs to start a day's work. The employer benefits with lower costs (specifically less overheads) and potentially from a more innovative workforce.

The communications infrastructure and social infrastructure of the internet may one day mean that the only people who go to a collective place of work are those who make physical things requiring a cooperative hands-on effort, or those who deliver services that require a physical interaction – such as dentistry or hairdressing. In the manufacturing industries, there may be even fewer manned manufacturing plants in the future than we currently imagine. Take the car manufacturing plant. Older depictions of the factory in the media sometimes show the factory's whistle blowing followed by a stream of people leaving the gates. A favourite car advert of one of the authors depicts the robots in the factory having a bit of fun with the paint brush while no one is looking. The person (rather than robot) actually having all the fun by making robots behave like human beings could be anywhere – in all probability as far away from his or her boss as possible. The auto industry has gone through a shift in the make-up and number of employees as a consequence of new production line technology. The (perhaps environmentally friendly) cars of the future could be made with some of the factory workers sitting anywhere in the world, and new designs could also come from anywhere in the world – especially from countries with a strong incentive to grow, such as China. Many modern factory employees are increasingly so-called 'knowledge workers'.

Many environmentally driven changes to the workplace could have unintended social consequences, and not all of them negative. If the split between tasks that must be done in a communal workplace or service-point changes, so that more tasks are done behind the scenes, the number of jobs that can be done flexibly could increase and this could result in more flexible working practices for many more people, levelling the virtual workplace for men and women during the child-bearing years of the latter. (This may be the flipside of the economic perspective above, which predicts long hours for some and part-time work for others – but, experience suggests, it may be positive only with proper regulation of pay and conditions.) Take food retailing as an example. At the checkout, an increasing number of the tills in some UK supermarkets are automatic. Currently, when something happens that requires assistance, people are on hand to help out. When the issue is to check that the purchaser of alcohol is above the age of 18, one asks why this manual check could not be done remotely by a team of people sharing shifts according to availability, leaving the small on-site team to deal with other aspects of customer service and potentially reducing commuter hours for the workforce in aggregate.

In a chapter of this length, it is impossible to explore such possibilities other than very superficially. The key point is that, because of new technologies under development and indeed already available, we will one day look back at the second decade of the 21st century and compare it to the way we spend our working lives now in sheer disbelief. Many of these changes are likely to be good for the environment, because they can be expected to reduce activities – such as commuting – that are driven by fossil fuel energy. But, if environmental considerations are not embedded throughout the entire manufacturing and service delivery chain, other aspects of environmental stewardship could end up being left to chance. Think, for instance, of an economist working at home in a draughty garret with oil fired central heating using data stored on servers powered with coal energy. Unless homes as well as the infrastructure supporting the economy are as energy and carbon efficient as possible, a shift to home-based working might not be as beneficial as it should be in environmental terms. Getting this right means getting *all* the ducks in a row, and there is much to lose by not doing so.

This section of the chapter brings us full circle back to the issue of education. The workforces in the growing knowledge-based industries of the future will need to be flexible and adaptable, learning as they go along. People who are fully engaged in the virtual as well as the physical leisure environment are likely to be honing some of the skills needed for the workplace. Those learning, training and working in the new information age are likely to have more fun in their free time. For those that do not adapt, for whatever reason, unemployment is likely to be an increasing risk. However, those that adapt will potentially bring with them the capacity to adapt to the need to use resources differently. What this all means, from the perspective of the environmental crunch, is that education and knowledge have never been more important.

As it happens, a good education has never been more important from the perspective of economics either, because, as discussed above, the long-term unemployed are disproportionately the low-skilled.

The financial credit crunch and the economics of leisure

And so, finally, we come to the topic of leisure. What does one do when one is not working or being educated? (Clearly, an economist has little direct personal experience of leisure. An economist's life is ceaseless toil and even something so mundane as shopping or visiting a bar is in fact nothing more than a fact-finding field trip into the real economy.)

Our leisure time is theoretically free from work or responsibility. Depending on how exhausted we are after the daily toil, commute, domestic duties and child care, we might want to spend it inactively – sleeping, lounging around or spending a few hours as the proverbial couch potato in front of the TV. However, many of us living in developed economies, defined as those in which relatively little time is needed to procure enough calories for subsistence, may have time and energy to spare for more energetic leisure activities on top of what we need to engage in to earn our keep and recover the energy to go back in the next day (such as doing the laundry and getting sufficient sleep).

Leisure time can thus perhaps be described as the absence of work, which means employment could be considered important to determining leisure – hence, the character of the Dowager Countess of Grantham, in the TV drama *Downton Abbey*, asking with clear confusion 'What is a weekend?' The financial credit crunch has two economic consequences on leisure. First, through work, it changes the amount of time available for leisure (and what one can do to occupy the leisure hours). Second, it changes the money available to be spent in the pursuit of leisure.

The amount of time available for leisure is, more or less, an inverse relationship to the amount of time dedicated to work, logically enough. For people in work, therefore, the amount of time available for leisure is likely to be reduced by the effects of the financial credit crunch. If working hours increased, or labour force participation rates increase, then the sum total of leisure time in a society will fall.

What this means is that leisure consumption is likely to be reduced, or potentially to be consumed in a more intense manner. If we take leisure in its very broadest sense – time not spent in work, education or asleep – then there is an obvious hierarchy of preference as to how one spends one's leisure time. Domestic housework may come low down on the list. Sport, watching television or reading insightful books on environmental economics may come higher up the hierarchy.

The consumer, who is more *leisure* time poor than before the financial credit crunch, therefore will try to minimize the time spent on non-work activities low down the list, and concentrate on what might be termed the 'higher pleasures'. (Higher pleasures in this sense can still be pretty base activities. It is just that any individual consumer gets more pleasure from them.) Labour-saving devices to reduce time spent on housework. Buying ready meals rather than preparing food from scratch. Shopping on the internet at work, rather than visiting physical stores. Each of these is a possible means of minimizing the amount of time spent on what any one individual may term a 'lower pleasure.' The environmental implications will, of course, vary. Internet shopping may be less environmentally damaging than physical shopping, depending on the physical infrastructure required to deliver the goods actually purchased in cyberspace. Cooking ready meals rather than preparing food at home is generally more resource-intensive.

The financial credit crunch does not render everyone leisure time poor, of course. For some, leisure rises to an infinite amount. The long-term unemployed effectively have unlimited leisure (in the sense that they have no paid employment). This does not mean, to adopt a somewhat archaic phrase, that they are 'enjoying their ease'. Money may not buy happiness, but it generally affords a lot of the leisure activities that exist. Restrictions on income, either as a result of unemployment, or as a result of the restrictions on household cash flow that arise from the financial credit crunch, will constrain the ability to consume leisure as they do with any other part of consumer spending.

Leisure costs money. This much is obvious. The financial credit crunch constrains leisure spending, as with anything else. However, the nature of the financial credit crunch offers a peculiar constraint on leisure spending. Leisure spending *can* be social – for example, in the case of shared activities. If there is a financial credit crunch, therefore, social leisure spending will tend to shift to what we might term 'the lowest common denominator' in a social group. Even if one can afford to undertake more expensive leisure-related activities, one may eschew them because one's friends cannot afford to join in. So, eating out declines, not because everyone is income- or credit-constrained, but because the friends one wishes to spend time with might be.

This process can be compounded by peer pressure. The media-constructed phrase 'staycationing' simply means taking a holiday at home. To one of the authors, in their youth, taking a holiday at home meant taking a holiday (with no need for a staycation qualifier) because foreign holidays were a rarity; and, for the other, camping was the means by which a rich tapestry of formative experience in foreign climes was frugally obtained. In the past two decades, hotel-based foreign holidays have become a norm. However, now that the financial credit crunch has hit, the fashionable trend is to be seen to be frugal in one's spending. The idea of foreign holidays has, if not declined absolutely, become less of an aspirational good. As we saw in the chapter on fast goods (Chapter 7), fashion can be a powerful force in adjusting consumer behaviour. Prior to the First World War, it was considered vulgar to have a sun tan (as it indicated that one was working class, labouring in the fields). From the 1920s, having a sun tan indicated wealth (affording the cost of a foreign holiday) and leisure (affording the time for a foreign holiday) and so became a status symbol. Perhaps, if ostentatious displays of wealth become sufficiently passé in a post financial credit crunch world, having a sun tan (and the foreign holiday that produces it) will fade. It may not hurt that the health risks associated with a sun tan (and indeed with the health of the environment) may also deter sun-seekers.

None of this means that leisure spending will necessarily decline absolutely. A population that believes it deserves 'a treat' for enduring the vicissitudes of the financial credit crunch is unlikely to deny itself leisure absolutely. What it does mean is that the patterns of leisure spending are likely to shift. One possible trend is that leisure becomes somewhat more insular. If ostentatious displays of wealth are frowned upon, then spending on leisure becomes a private affair. If large social interactions (in the real world) become more difficult to arrange because of the 'lowest common denominator' problem, then private rather than public occupation of leisure becomes normal. Home consumption of beer rises, rural pubs close.

Environmental perspective on leisure, post financial credit crunch

One key question for this chapter is whether the potential volume constraints inherent in environmental limits (and related social limits) such as congestion will increasingly restrict how we spend our leisure time. Taking a cue from the work of Fred Hirsch writing in 1977, many of the things we do in our spare time can be described as 'positional' goods. This term refers to goods or services that are 'scarce in some absolute or socially imposed sense' or 'subject to congestion in some absolute or socially imposed sense' (Hirsch, 1977, p27). Many leisure activities fit this description. We do not have unlimited time for leisure (or as discussed above, if we do, we may not have the wherewithal to enjoy it post credit crunch); and when we do indulge the leisure activities typically enjoyed in the 21st century, we often find that the facilities and infrastructure through which they are made available are congested. The presence of environmental constraints exacerbates the scarcity problem – whether through actual environmental impacts (such as polluted beaches rendering maritime leisure unattractive or impossible) or through socially imposed constraints (such as the imposition of carbon costs or direct limits on numbers).

Traditionally, many leisure activities have entailed physically moving to the appropriate place – sports facility, eating or drinking establishment, cinema, shopping centre or airport. All of these activities have their particular environmental impact – the venues listed are in approximate ascending order of the carbon footprint of each of the activities involved. As we have discussed in previous chapters, efforts are underway to try and de-emphasize the environmental footprint of such activities, but because each of them depends on a specific piece of infrastructure this can take a very long time. In the medium term, the buildings and transport facilities that support these activities will (in response to a widening body of environmental regulation) increasingly be built in such a way as to minimize their environmental footprint, both during the building of them and then, afterwards, in their usage by consumers.

The problem is what to do with existing infrastructures, which tend to shape the environmental footprint of our leisure time in ways we cannot control. One of the authors of this book, as a student, accustomed to wearing out shoe-leather to get from one place to another as a tourist in places such as Madrid and Paris, remembers going to the US and being stopped by the police as a potential runaway juvenile for doing something so unusual as walking from one place to another in a major US city. A postcard home described this benign encounter with the law and continued: 'No one walks anywhere – you even need a car to get takeaway food.' As this illustrates, how cities are designed shapes the way leisure time is spent – as well as determining its environmental footprint (and, come to that, its impact on health too).

The existence of a global transport infrastructure means that currently it is both cheap and easy to travel physically to distant locations and this means that leisure travel entails growing environmental costs alongside the many benefits it brings with it. Starting first with the benefits, increased travel opens up cultural exchange, therefore arguably increasing other forms of exchange (such as trade) and reducing (through increased understanding) geopolitical friction. Thus, cross-border travel is something the European Union (EU) is particularly keen on because it aids unification – but, of course, the increase in air miles means increased greenhouse gas emissions. These points, in parallel, raise a number of very interesting questions in the area of leisure and culture, from an environmental perspective.

One of the benefits of the new possibilities for the consumer to go beyond their immediate surroundings is cultural. It is now much easier than it used to be to see the physical effects of agriculture upon rainforests or the overuse of water on the Gobi desert, for example, and the signs are that this improved access to information is causing some consumers to change their behaviour. However, the flipside of this is that the 'because I deserve it' culture of consumer spending in developed countries could easily be adopted worldwide. When information in any part of the world is immediately global, the consumer culture of individual countries can have an influence that goes far beyond their own borders, with both positive and negative effects. This is something the financial credit crunch is likely to have little direct influence on. However, the message must be 'nil desperandum' for things can change in unexpected ways.

The pre-internet generation – those who remember the days when arcane communication media such as the telegram was the only speedy way of communicating in writing for most consumers outside the office – look back to the mid-1990s and compare it to the way we spend an increasing proportion of our leisure time now in astonishment. By extrapolation, an optimistic perspective would suggest that the significant change needed in our relationship with the environment can happen however impossible this may seem.

Consider the internet. The average young person now spends at least as much time in front of the computer networking with friends in cyberspace as networking with friends in terrestrial space. The nuclear family (two parents, two children) of the post-war era replaced the extended family (grandparents, aunts and uncles living in relatively close proximity) of the pre-war era. Now, the nuclear family seems to be being superseded by a virtual family – interacting via social networking sites and constantly aware of what each other member is doing. The family members may or may not be related by blood, but the community concept is there in a virtual sense, just as it was two generations ago in a physical sense.

Not only that, the presence of the so-called 'silver surfer' pops up regularly in the media, suggesting that the internet is changing things for the full range of generations. People can live double or even multiple lives, as avatars[4] in a virtual economy and physical people in the conventional economy. As the generational cohorts march steadily through time, the new media will inevitably change the way leisure time is spent; one question being whether time spent in the context of the virtual economy has a smaller or greater environmental footprint than time spent in the real economy. Without energy and equipment from the real world, the virtual economy could not exist (electricity drives data storage and transmission, and the durables through which connections are made have their own footprint). Nevertheless, a trip to inspect the rainforest taken in cyberspace might be more desirable to the consumer (on the basis of its significantly lower cost in time, physical wear and tear, money and carbon emissions) than the physical trip, and the individual's grasp of the *facts* is unlikely to be diminished. (However, if the cyber-traveller never gets out into their local countryside, this may, on the contrary, translate to a reduced understanding of the importance of biodiversity however closely rainforest damage can be inspected by virtual means.)

As we write, increasing numbers cannot imagine life without the 'net'. Shopping, travel arrangements, entertainment, auctioneering, information search, academic research, training and education, and personal administration such as tax returns all have a significant and growing virtual service infrastructure in place. It is worth pausing to reflect briefly on the obvious: the invention of the internet was made possible by the ingenuity of people who were highly trained in mathematics and computer programming, as well as perhaps other social sciences. Without education, it is unlikely to have been possible – we do not believe that a group of monkeys, handed computers and left to type at random would have arrived at the programming that made the internet possible any more than they could create the works of William Shakespeare.

The internet, at the stage of development it is reaching at the beginning of the second decade of the 21st century, is democratizing information. The internet is therefore also spreading an economic and environmental culture, defined as the set of customs, ideas and social behaviour observable in the context of economic activity. The impact is also instrumental in allowing economic and environmental culture to have an impact upon the environment. This can have unintended consequences. In their paper entitled 'Why do humans reason?', cognitive scientist Dan Sperber and philosopher Hugo Mercier ask 'What if we evolved the capacity to reason not to get closer to the truth but to persuade others (and ourselves) of viewpoints, regardless of their relation to the truth?' (in Burkeman, 2010, p115). What happens to the experience of the cyberspace traveller in the rainforest mentioned above if someone somewhere were to 'adapt' some of the content on Google Earth? Note that we are not suggesting this is the case – but, the global reach of something such as WikiLeaks raises all sorts of questions about the integrity of the information found online. Internet users need to use their judgement – and this (once more) implies the need for an all-round education.

At the most basic level, the new and rapidly expanding resource of the huge social network provided by the internet can only be enjoyed by the people who have access to it if they have, as a minimum, a good high school education. It is only likely to be used to the best advantage by people who have received an education sufficient to allow them to use their own powers to the full in exploiting it or contributing to it. The need to question information (in case it turns out to be propaganda or 'spin') is nothing new. The most topical example in the field of the environment is the idea that greenhouse gas data could be used to predict the future level of global warming with precision. This is based on the more generally held belief in the myth that statistical relationships are precise and certain. Rising temperatures are observable in fact, as is the rising level of greenhouse gases, but what this means for the climate (and the environment) can only be predicted with a margin of error. Unfortunately, this subtle point (but not the statistical myth) was the baby that 'Climategate' threw out with the bathwater. This example takes us right to the heart of the reason why a good level of education for as many as possible – with the goal of democratizing the ability to absorb, assess and, where needs be, challenge new and old information – is absolutely critical; the ability to absorb and critically assess information will directly affect the relationship that the human race has with the environment in the long term.

It also matters enormously in an absolute sense. Going back to John Kay's interview with Peter Day on BBC Radio 4 on 25 November 2010, he pointed out that what really matters is the growth in productivity that comes from human ingenuity – a resource that can reasonably be described as unlimited. As Kay put it, we need to divert human ingenuity into giving us in the long run 'things we will think we actually want', such as iPods, iPads, clean energy, effective medical treatment or Harry Potter, 'rather than into giving us things we wish we hadn't had, like sub-prime lending'. And, indeed, like the environmental crunch this book is concerned with.

It is stating the obvious to say that education shapes leisure. Before the advent of literacy, people could not entertain themselves (or indeed seek out entertainment) if to do so they needed to read or write. However, the interaction between education and leisure is a two-way street. As Dr Johnson wrote: 'All intellectual improvement arises from leisure.' Many of the inventions that have led to technological change arrived because their inventors were playing with ideas. Archimedes is certainly not the only person to have come up with a great idea in the bath. The lives lived by avatars (see above) are just one example of human creativity on the internet – they would not be possible without the presence of a numerate and literate cohort of people at work and, of course, at leisure.

Conclusion: Learning, working and playing with two credit crunches

The modern world is, in some ways, no more subtle or complex than it was several hundred years ago. When trade expanded in the Elizabethan era, highly educated people who had the advanced rhetorical skills needed to communicate effectively in politics and those who could use mathematics and languages in the expansionary traded goods environment were eagerly sought after. In parallel, more and more had access to the rich literature, art and music of the day. Then, as now, education, work and leisure-time activities had the potential to work as a rich social continuum. The social and environmental problems facing us in the modern era are different to those facing our forebears only in scope – environmental impacts in the pre-modern age were primarily local. Now, what we do at work and at leisure in one country can have significant environmental impacts in many others. This is partly because of the physical trade associated with economic activity but also because modern, developed and highly networked economies have the power to influence work and leisure practices – and therefore the environment – worldwide in both good and bad ways.

Notes

1 A *Times* online article (8 June 2010) cited a 2010 report by the Central Council of Physical Recreation, as suggesting that the UK had lost 6000 playing fields since 1992, leaving just 20,000 in place. However, initiatives appear to be in place to reverse this decline, sparked by the approach of the 2012 Olympics. See www.timesonline.co.uk/tol/news/uk/article7145758.ece.

2 Jensen and Meckling (1994) consider the example of life-time employment, which had become the norm for several decades in the US, Europe and Japan, but then gradually had to be abandoned (beginning in the US) as the pressures of global competition increased and firms could no longer hold to this informal contract.

3 See CRC Energy Efficiency Scheme, as described at www.decc.gov.uk/en/content/cms/what_we_do/lc_uk/crc/crc.aspx.
4 An avatar could be described as an internet alter ego, sent out to socialize in cyberspace on behalf of its creator. It often takes the form of an image in 2D or 3D. The more sophisticated versions could become environmentalists. This of course would make them hard for economists to keep up with (because economists are firmly based in the real world).

References

Burkeman, O. (2010) 'This column will change your life: When reasoning goes out of the window', *Guardian Weekend Magazine*, 27 November
Hirsch, F. (1977) *Social Limits to Growth*, Routledge and Kegan Paul, London.
Jensen, M. C. and Meckling, W. H. (1994) 'The nature of man', *Journal of Applied Corporate Finance*, vol 7, no 2, pp4–19

CONCLUSION

Change: The Consumer and the Twin Credit Crunches

In the struggle for survival, the fittest win out at the expense of their rivals because they succeed in adapting themselves best to their environment. (Charles Darwin, *On the Origin of Species*, 1859)

The twin credit crunches that have run through this book are complex subjects. Volumes can (and doubtless will) be written about the many aspects of both the financial and the environmental credit crunch. However, the underlying concepts of a credit crunch are simple. Let us start with the hierarchy of definitions that has brought us to this point:

- Credit is simply a mechanism that allows us to consume tomorrow's resources today;
- Credit necessarily means that consumers accept a lower standard of living in the tomorrow than they would otherwise have enjoyed, in order to enjoy a higher standard of living today;
- The concept of credit can therefore be applied to financial credit or to environmental credit. Financial credit means consumers borrow against tomorrow's income to consume today. Environmental credit means consumers consume finite resources today, which means they cannot consume these resources in the future;
- Finally, a credit crunch is an event that gives lie to the proverb 'tomorrow never comes'. A credit crunch is when credit is no longer available and when past credit borrowed must be repaid. A credit crunch is when the consumer must start living with tomorrow's standard of living.

The financial credit crunch of 2008 onwards was a realization that (within Organisation for Economic Co-operation and Development (OECD) economies at least) the consumer could not continue to live beyond their means at the pace that they had been doing. The environmental credit crunch is recognition that the consumer cannot live beyond their means at the pace that they have been doing. The financial credit crunch is immediate, visible and (in terms of its personal consequences) reasonably well-understood.

The environmental credit crunch is more slow-moving, and relatively visible because it is the subject of a great deal of argument (even among the expert community). However, in the scheme of things, it may turn out to be a far bigger problem than the financial crunch currently underway. Taking a cue from the Global Footprint Network:

> *Humanity relies on ecosystem products and services including resources, waste absorptive capacity, and space to host urban infrastructure. Environmental changes such as deforestation, collapsing fisheries, and carbon dioxide (CO_2) accumulation in the atmosphere indicate that human demand may well have exceeded the regenerative and absorptive capacity of the biosphere. Careful management of human interaction with the biosphere is essential [if an environmental credit crunch is to be avoided].* (Global Footprint Network, 2010, p1)

Recapping on where we began, in this book we ask whether the financial crunch might potentially trigger changes that could lead to the avoidance of an environmental crunch. The steady trend towards an earlier and earlier Earth Overshoot Day depicted in the chart in the Preface (based on Global Footprint Network data) is ominous, suggesting that an environmental credit crunch may be inevitable unless *something* happens to change the current trend. Is the financial credit crunch that *something*?

The number of 'planets'-worth of environmental resource required to support the number of human beings on the planet, as well as the way human beings in aggregate live their lives, is steadily growing, as reflected in the steady shift forwards of Earth Overshoot Day clearly visible on the chart. We moved into the 'red' from an 'environmental income' perspective in the mid-1970s. Since then, Earth Overshoot Day has moved forwards by about four days a year on average and in 2010 arrived on 21 August. This means that one and a half planets would be required to generate the so-called ecosystem services used in the course of 2010.

In 2010, we were living on the 'never never' for the last 132 days of the year. In the sense that we are 'only' in the red for a few months, and have 'only' been in the red for a portion of each of the past 34 years (a short period of time in the history of civilization), getting out of the 'red' back into the 'green' is not impossible. Note, however, that in environmental terms, a complete environmental deleveraging not only requires Earth Overshoot Day to fall on 31 December each year, but also for damage to the environmental 'balance sheet' wrought by previous years in the red to be repaired. As a first step, getting back into the 'green' on a running annual basis is not an impossible target in light of the relatively short time we have spent in the 'red'. But, as everyone who has been in debt knows, the longer debt persists and the larger it gets, the harder it becomes to get back into the black, and this is why there can be no delay.

In a financial sense, deleverage requires a multiplicity of small changes to scale back excess spending across a range of activities, as well as to increase income from employment (and other sources). Similarly, in an environmental sense, there is no single solution to the conundrum of the environmental credit crunch. At the margin, we expect the financial credit crunch to bring about relevant changes in behaviour. However, some of the environmentally helpful changes – those that bring about a reduction in consumption – will often only be temporary. Other changes (such as trading down in response to financial constraints) may not help and indeed may be out and out damaging to the environment if for instance 'trading down' fast-moving goods in terms of quality and price should also result in a down trade in environmental terms, in relation to the materials used or the more rapid disposal of such products. Moreover, the financial credit crunch will tend to constrain the investment spending needed to move to an environmentally friendly way of life, on all fronts – be it funded by government support, private sector investment, capital markets, tax revenues or direct consumer spending. On the other hand, there are no limits to what human ingenuity is capable of and the consumer, who is also a voter and a taxpayer, actually has significant latent power to bring about change.

The coinciding of these two crunches – the financial and the environmental – is a global challenge. What can we say about how both of these credit crunches have developed so far, and what lessons can we draw for how policy and economies need to develop in the future?

The financial credit crunch

The financial credit crunch is primarily economic. That is not to say that there are not economic aspects to the environmental credit crunch – there are of course. There is virtually no aspect of human existence that an economist cannot pontificate on, given the chance. However, with the financial credit crunch, the impact is most directly seen in the economy.

So what has the financial credit crunch meant? Again, the basic issues are simple. There is less financial credit available (the ability to borrow, or 'leverage', is reduced). Financial credit is also likely to be more expensive. One can also say that, as a result of these two changes to financial credit, the trend rate of growth in an economy is likely to be lower than it has been in the past.

So far, so simple. The problem is not what we might term as the 'first-order' effects of the financial credit crunch. The problem is the way in which consumers react to those first-order changes, and then the way in which politicians and policymakers react to the consumers' reactions. Here things become a lot more complex, very rapidly. This complexity arises because of two economic facts:

- The financial credit crunch changes things;
- People do not like change.

The past 20–30 years have seen change as well, of course, but it has been gradual and (generally) in one direction. Consumers have experienced a rising standard of living. They have been able to buy more and more stuff. They have been able to ignore issues such as growing income inequality by taking regular, hallucinogenic doses of consumer credit. This is the change consumers can believe in (because it is nice change, while it lasts).

Now that financial credit is denied and past credit bills are falling due, the consumer has to face up to the unsustainable nature of their lifestyle. They are asked (and not terribly politely) to change their spending levels. They are asked to change the total amount of goods that they purchase. Through the lower rate of growth, they are asked to change their expectations for future income growth and future standards of living.

Confronted by this shift, the consumer will attempt to rebel. As every politician seeking re-election knows, the consumer does not readily accept that their standard of living must fall (or that the rate of growth in their future standard of living must be constrained). The financial credit contract of the 1990s and early years of this century was that consumers would accept a lower standard of living in the future, in order to enjoy the good life in the present. Now that the future is the present, the consumer is trying to default or renegotiate that contract. Consumers are clinging to the material possessions of the past and trying to hang on to what they have. Politicians are responding to this by trying to preserve what already exists, economically speaking. The problem we have tried to highlight in this book is that preserving the economic status quo is not consistent with the environmental credit crunch.

The main economic consequences of the financial credit crunch can be summarized as affecting consumer spending, employment, investment and growth. They are all interrelated of course. But these tangled threads are woven into a process of change.

The financial credit crunch and consumer spending

By definition, a financial credit crunch will reduce first the amount, then the growth of, consumer spending. When credit is available, consumers can spend more than they earn. Without credit, consumers can spend what they earn. With a full financial credit crunch, the consumer can spend less than they earn – because they are required to repay debt.

This slower consumption inevitably means fewer resources are consumed. However, consumers' aversion to change means that they will seek ways to preserve their standard of living (especially the outward display of that standard of living), even with a more restricted family budget. 'Trading down' is a theme that has run through the economic sections of this book. If something can be purchased more cheaply (in economic, if not environmental terms), then it is seized upon as a means of maintaining a material standard without the inconvenience of having to pay more for it.

Service sector spending also suffers for the same reason. Service sector spending is vulnerable because it is less essential in a materialistic society (there is no visible 'stuff' to display from spending on a service). Why pay someone to cook a steak for you, if you can cook it at home?

Thus, consumer spending habits change. The more enduring the financial credit crunch is, the more enduring these changes are likely to be. Consumers are likely to respond by cutting back on service spending, seeking cheaper goods and possibly trying to extend the life of existing possessions through repair rather than replacement.

The financial credit crunch and employment

If consumer spending patterns are shifting, there will be a shift in the employment patterns that go to support that consumer spending. Service sector jobs become vulnerable if consumer spending growth is less reliable. Manufacturing sector jobs (in the OECD) potentially become vulnerable as consumers seek cheaper goods (thus cheaper labour) from overseas.

Moreover, because consumers are credit constrained, we are likely to see an increase in part-time working. If consumers cannot spend more than they earn, but want 'stuff' they cannot afford, the options available are crime (a route that will be taken by some in an economic downturn) or increasing household income. Add the veneer of job insecurity for many households and there will be a desire to increase participation rates (which generally means female participation rates in the OECD countries).

This creates a complex labour market reaction. At one level, there is likely to be an increase in unemployment (slower growth will tend to do that). Due to the length of the economic downturn, and the shift in spending patterns, there is quite likely to be a permanent increase in unemployment (a rise in the 'natural rate of unemployment'). Long-term unemployment will rise most particularly among the low-skilled workers within the OECD. This in turn creates a challenge in terms of income inequality. As low-skilled workers tend to coincide with low-income workers, the change in employment patterns is likely to threaten an increase in income inequality. This may not be across the spectrum, but it is likely to occur between the lower-income groups and the rest of society.

At the same time, there is also likely to be an increase in employment among middle-income households. The key word is households, because what we are talking about here is an increase in employment participation (particularly among the middle classes), as households try to maintain the materialistic status quo. This may increase the material (or financial) income of the household, but it also makes the household increasingly time poor.

These changes in employment patterns obviously feed back into consumer patterns. Change breeds change in economics. However, as we have also seen, change will be resisted. Consumers will not necessarily like the change of job insecurity or (fairly obviously) higher permanent unemployment levels. As a result, politicians will respond to the consumers' disquiet.

Nothing tends to motivate politicians more than unemployment. It is a spur to action because their jobs are at risk if it is allowed to persist. As a result, the political response to the financial credit crunch is likely to be to seek employment security (for those in work) and job creation (for those out of work). However, politicians will move conservatively, seeking to preserve (or 'conserve') jobs that already exist but that are at risk, even if those jobs are economically or environmentally redundant. If job insecurity is a consequence of change, then security can be achieved (relatively simply) by turning back that change. This does not mean reversing the financial credit crunch (that is too expensive for a resource-constrained government to contemplate). It means using subsidy or regulation to distort the market so as to preserve the past.

Subsidy distortions may perpetuate economically and environmentally outdated industries. Regulation is also distorting, and in particular (through the regulation of global trade, otherwise known as protectionism) may spread the costs of the financial credit crunch further around the world. Economists tend to recoil from both trends. Politicians require strong willpower to reject the siren calls, as they may provide a temporary illusion of improving economics.

When we turn to job creation, there is more scope for some positive thinking. Politicians could create jobs that are economically and environmentally positive. There has been much talk of 'green industry' as a solution to the current crisis, for instance. While this is laudable as an aim, and a positive response to the financial and environmental credit crunch, the economics of this does not necessarily stack up. The unemployment problem is likely to focus on lower-skilled workers. Many green jobs (for instance those in the energy sector) are unlikely to be low skilled. At the very least, green jobs will require new training (self-evidently; we have not been terribly 'green' in recent years and so some kind of retraining will be required). Governments therefore face a challenge in matching skills with future objectives – and it is a daunting challenge.

The financial credit crunch and investment

The third strand to run through the economics parts of this book is the investment consequences of the financial credit crunch. Investment is maximized when consumers, companies and governments can borrow. Investment tends to produce returns over a long period of time. Financial credit allows costs to be spread out over a long period of time. It is a great match.

The financial credit crunch limits investment. There is no possibility of questioning this – it is absolutely the most critical issue for economists, and economists (of course) are never wrong. The limitations of investment are widespread.

Limiting investment means limited investment for households. For some, this may mean keeping existing durable goods for longer through repair and maintenance. It may also suggest deferring the upgrading of consumer durable goods for a longer period. One may endure higher economic and environmental running costs in order to defer the capital cost of 'investing'. That could be either because the capital costs have gone up (borrowing is more expensive) or because the borrowing to supply the finance for capital investment is now denied.

This same process is repeated, with varying degrees of severity, for corporate and government investors. Existing infrastructure is kept in place for longer than would have been the case when credit was freely available. This is one of the driving factors behind reduced growth, of course. Reducing investment will tend to reduce growth directly (the act of investing is growth positive) but also through the limits lower investment will tend to put on productivity. Productivity is correlated with investment spending, and by undermining productivity there will be a lower trend rate of economic growth in the economy.

We should not regard investment as being all about investment in machinery, equipment, durable goods and infrastructure. This is a big part of the investment story to be sure. But, just as the world is divided by economists into labour and capital, so we can divide investment into labour and capital as well. Investment in labour (education and training) is also threatened by the financial credit crunch. That can have some widespread implications for future economic development and feeds back into the employment position, and through that back into the consumer spending position.

The financial credit crunch and growth

Economic growth is about making more stuff, year after year. If supply is growing, then so too must demand (ultimately). Economic growth is about consuming more, year after year. Economic growth therefore has environmental consequences, as it necessitates the consumption of renewable and finite resources (and renewable resources that through their overconsumption may become finite). Economic growth also generates humanitarian credit consequences (growth raises standards of living, after all).

Credit enables consumers to consume more. Financial credit crunches are usually about consuming less or consuming at a slower rate. What is different about this particular financial credit crunch is the possibility of significant changes in behaviour (such as increasing consumer sensitivity to the costs associated with waste highlighted in the energy chapter or the possibility that an energy crunch would go a long way towards prompting greater energy efficiency commented on in the housing chapter). The effect of such behavioural change, triggered by the catalyst of the credit crunch, could be to bring about a structural break in the patterns of consumption that would (by bringing about related structural breaks in growth trends) help push Earth Overshoot Day back towards the end of the year.

Most reputable economists would agree that the trend pace of economic growth has slowed because of the way consumption, labour markets and investment have conspired to change. This makes the allocation of finite resources amongst infinite desires (the raison d'être of the economist) more difficult. Managing change is always easier in a more rapidly expanding economy. If the economic 'cake' is growing in size, then it is easier to adjust the portions. No one gets a smaller slice of the cake, it is just that some get even larger slices than they would otherwise. However, when growth slows (or stops), managing change becomes a problem. If someone has a larger slice, then someone else must accept a smaller slice of cake. And no one likes a smaller slice of the cake.

The economics of the financial credit crunch

The consequences of the financial credit crunch are at once simple, and disarmingly complex. There is less financial credit available. That much is self-evident. The very name 'financial credit crunch' gives away the conclusion. However, the complexities that arise from the financial credit crunch are many. Change is coming. These changes will impact society, the economy and inevitably the environment.

The financial credit crunch does not take place in a vacuum. It takes place in a physical environment that has seen consumers charging to their environmental credit cards with reckless abandon. The use of environmental credit can be said to precede the use of financial credit (the use of finite resources has been going on for some time). In contrast, the denouement for financial credit (the financial credit crunch) precedes the environmental credit crunch – but not by much. The way we live now – and the way that our style of living is changing as a result of the financial credit crunch – will determine how quickly the environmental bailiffs descend to demand recompense for the environmental debt that consumers have built up.

The environmental credit crunch

As the nine chapters of this book suggest, a range of things could be done to avoid the arrival of the bailiffs; the emphasis being on working on as many different directions as possible – because dealing with the environmental crunch must be a matter of trial and error in light of its complexity. This latter point is discussed in the context of economics by a well-known writer. (Even an environmentalist has to concede that economics does have its uses.) John Kay commented:

> *Markets are not about harnessing greed ... Market economies succeeded because they established pluralism and put in mechanisms which solved, or at least reduced, problems of incentive compatibility. The twin pillars of disciplined pluralism and incentive compatibility should support economic policy.* (Kay, 2003, pp355–356)

So, to plagiarize Lenin, 'What is to be done?' Below, we offer some suggestions for each of the sectors reviewed above, based on the relevant chapter, noting (in the spirit of disciplined pluralism) that this is certainly not an exhaustive list of possibilities.

TABLE C.1 Action points for food

Action point	*Financial credit crunch to trigger action?*	*Points to note*
Reduce food waste.	Limited reaction from consumers likely; a more significant reduction possible.	Depends on whether food retailing practices change.
Change eating patterns – for example, less meat, fewer calories.	Possibly.	Depends on pricing and changes in culture.
Establish sustainable farming practices in developing countries as an integral part of building food production capacity.	No.	Requires a political solution.
Reflect the environmental costs of food miles in food freight.	Sensitivity to environmental costs (where they are reflected) is likely to increase, but no *new* trend to reflect environmental costs is expected.	Likely to require regulatory intervention, e.g. extension of GHG regulation such as the European Emissions Trading Scheme.
Stimulate research and development (R&D) in sustainable agriculture.	Likely to be made worse.	Could be sparked within the corporate sector by the need to control supply-chain risk.
Make it easier for the harried urban worker and two-income family to access fresh food without planning ahead.	No, and possibly made worse (through increasing time poverty).	The stimulation of social enterprise (by government or private sector subsidy) may be necessary to make this possible.
Stimulate R&D in environmentally friendly approaches to food production.	Made worse.	Subsidy may be needed – or alternatively a healthy academic sector.
Protect biodiversity.	No.	An insurance approach to biodiversity may be needed. Funding for research also.
Persist with attempts at global political collaboration (such as Kyoto).	No. To the extent that protectionism increases, and undermines multilateralism, may be made worse.	Solutions to the environmental crunch problem for food require a pluralistic approach.

Food

Few of the things that could be done to render food production more sustainable from an environmental perspective will be triggered by the financial crunch alone. As Table C.1 suggests, a sweeping change in behaviour is needed and this will need to come from 'policy' action and innovation on the part of government, the private sector, individual consumers as a collective and academia.

Water

The financial credit crunch will only help bring about changes to the impact of human beings on the ecosystem through its usage of water if it happens to be the catalyst for a major cultural change.

TABLE C.2 Action points for water

Action point	*Financial credit crunch to trigger action?*	*Points to note*
Create incentive for the protection of the natural water infrastructure.	No.	International political agreement required.
Improve water efficiency in agriculture.	No, may make it worse through the reduction in credit for infrastructure spending.	Would be helped by an appropriate pricing mechanism, infrastructure investment and also R&D in agriculture.
Improve water efficiency in the industrial sector.	Possibly – a balance of running costs and infrastructure costs.	In some countries water is priced to allow economic forces to work to the advantage of the environment.
Improve water efficiency on the part of consumers.	No.	Consumers need more information about their domestic usage, and a better household water infrastructure, to control water usage.
Change eating habits.	Unlikely.	If water-intensive foodstuffs were more expensive, the economics of the crunch could affect eating patterns in a way that would help the environment.
Virtual trade in water to reflect water embedded in products that cross national borders.	Likely to reduce through trade protectionism (especially in agriculture).	Further liberalization of international trade, and most particularly trade in agriculture, is necessary to benefit from the trade in virtual water. This seems to be a receding prospect in the financial credit crunch environment.

Energy

The scope of the financial credit crunch to deliver a more environmentally friendly use of energy might lie (to an extent) in financial constraints, as people seek to save money by reducing the number of car journeys, or turning down the heating thermostat. However, these changes may not endure (President Carter and his beige cardigan serves as a reminder to us all). Energy use could rebound with the economy, although the slower-trend rate of growth suggests that it will not rebound as vigorously as past demand. Investment in energy R&D, clean energy and a better infrastructure is likely to be held back by the financial credit crunch, which could therefore have adverse environmental impacts in the long run. If productivity became less narrowly defined, so as to incorporate the environmental impacts of economic activity, this could change. The question is whether the financial crunch will eventually turn out to be enough of a shock to act as a catalyst for such a major shift in thinking worldwide. As things stand, this looks unlikely.

TABLE C.3 Action points for energy

Action point	*Financial credit crunch to trigger action?*	*Points to note*
Energy efficiency.	To an extent.	In the short run, consumers may reduce energy use but structural improvements may be held back by the crunch.
Energy substitution (green energy for alternative energy).	No.	The financial crunch will act as a constraint on all investment.
Investment in energy R&D.	Likely to reduce as government budgets are constrained.	Energy R&D tends to need government subsidy.
Investment in clean energy infrastructure.	Likely to reduce as government budgets are constrained.	As above.
Shift away from a narrow definition of productivity to a 'greener' version of gross domestic product (GDP).	Not in normal circumstances.	If a large enough shock to the system, the crunch could trigger the beginnings of change.
Create new, specialist credit markets in association with clean energy.	No.	Financial innovation may be held back by a harder regulatory regime for banks.

Infrastructure

The natural infrastructures of the planet play a regulating function, and the built infrastructure does exactly the same thing: how much the infrastructure grows has a bearing on how much volume can go through it. The financial credit crunch is mostly powerless to help reduce the environmental impact of the built infrastructure, on the basis that a downturn in activity will do nothing to bring about structural change. Only a redesign of large segments of the economy – working patterns, living patterns – could do this. In the shorter term, rationing of use could be environmentally helpful, but permanent changes in this direction are unlikely to be induced by the credit crunch, therefore government policy and cultural change will be required.

TABLE C.4 Action points for infrastructure

Action point	*Financial credit crunch to trigger action?*	*Points to note*
Take care of natural infrastructures impacted by human activity.	No.	Requires regulation by government or self-regulation by the private sector.
Build any new infrastructure with regards to the environment.	No.	Capital constraints and education limits will tend to impede innovation.
Change the way existing infrastructure is used, or adapt it with the same effect.	No.	However, financial constraints could temporarily reduce volumes of usage.
Change the way sectors that use infrastructure function.	No.	A downturn in activity could have short-run environmental benefits.

Housing

The design of homes, the urban environment and indeed the infrastructure that supports them offer significant scope to improve the environmental footprint of the housing sector. The barriers to change are, however, considerable. The pre-existing housing stock, together with the living and working patterns that shape it, will slow the pace of change. Non-structural changes (such as the widespread adoption of metering energy and water usage, and refurbishment incorporating efficiency improvements) require investment and therefore may not be helped by the financial crunch. On the other hand, this field could be a source of new jobs, with a modicum of regulatory guidance.

TABLE C.5 Action points for housing

Action point	*Financial credit crunch to trigger action?*	*Points to note*
Change the way homes are designed.	Not directly.	Government regulation in combination with a competitive private sector could deliver this. It is possible that increased sensitivity to efficient housing (from constrained consumers), combined with a more competitive housing market, will lead to builders using some environmental design as a unique selling point for new property.
Change the approach to urban design.	No.	Requires government regulation.
Changes to infrastructure.	No.	Requires long-range planning and changes elsewhere in the economy.

Durable goods

The replacement cycle for durables basically needs to slow down for the environment to be helped. The financial crunch may achieve this in the short run, but what is really needed is structural change to make this change stick. Some environmentally protective changes to the way goods are made (the avoidance of designed-in obsolescence, the provision of enough environmental information to facilitate consumer choice, a shift to modular manufacture) could offer scope for firms to cut costs, and therefore could be encouraged by the crunch. Through its shock value, the crunch could also help trigger cultural change, helping environmental credentials to deliver brand value.

TABLE C.6 Action points for durable goods

Action point	*Financial credit crunch to trigger action?*	*Points to note*
Reduce the frequency with which durable goods are replaced.	Yes.	This is likely to be most pronounced in the depths of the financial credit crunch. However, the slowdown in trend growth rates may lead to a slower pace of consumption of durable goods (i.e. less frequent replacement).
Change manufacturing approach – avoid designed-in obsolescence, give enough environmental information to facilitate consumer choice, shift to modular manufacture.	Possibly.	Modular manufacture can deliver efficiency savings, therefore cutting costs for the manufacturer. Repair and maintenance services can be a revenue source for firms.
Redesign durables for environmental efficiency in use.	On balance, probably not.	Efficient white goods are available – this therefore depends on consumer behaviour.
Cultural change – environmental credentials become (a component of) brand value.	Possibly.	The consumer has latent power to bring this about. To the extent that frugality becomes fashionable, it may have environmental consequences.

Fast goods

Many fast goods used to be durables. If they were re-re-categorized this could be environmentally helpful, but the financial crunch in isolation is unlikely to be enough to achieve this. As we said above, environmental considerations need to become so much a part of branding that designing a new product without taking the environment into account, or indeed purchasing a new fast good without doing the same thing, would become reprehensible.

TABLE C.7 Action points for fast goods

Action point	*Financial credit crunch to trigger action?*	*Points to note*
Return fast goods to the durables basket.	Yes and no.	Goods may be repaired (good) or people may trade down (potentially negative).
Redesign and market goods to change the way they are made and used.	Unlikely.	Firms need a premium to cover the cost of changing products. Only if redesign also reduces manufacturing costs can the crunch trigger such change.
Changes in fashion and/or culture.	To an extent.	The consumer has considerable power to influence firms to deliver the desired product.

Health

Changes to the way we live our lives that would be good for our health would also be good for the environment. Therefore, making changes with a view to pushing out Earth Overshoot Day would have the positive externality of improving health. (Health happens to be fundamental to economic performance.)

TABLE C.8 Action points for health

Action point	*Financial credit crunch to trigger action?*	*Points to note*
Improve diets – reduce overconsumption of calories in developed economies.	Unlikely to take place on a significantly large scale.	This would be simultaneously good for the environment and good for health.
Design the urban environment with health in mind.	No.	As above.
Incorporate health considerations in work.	No.	As above.
Increase R&D into the relationship between the environment and health.	No, may be made worse by financial credit constraints.	Would require regulation or support, or education spending. It is possible that existing education (the current stock of knowledge) will produce breakthrough research as a random act of chance, but the probability of further insight diminishes with declining funding.
Change the way productivity is defined.	Maybe.	If the financial credit crunch provides a sufficiently powerful shock, maybe.

Education, work and leisure

The relationships between work and the environment, and leisure and the environment are largely explained by the preceding chapters; and the main conclusions (that significant potential benefits to the consumer could arise from the incorporation of environmental considerations into work and leisure) apply here too. As above, we find that the financial crunch in isolation is unlikely to trigger such changes, so there is no need for a table of action points here, for these are mostly covered above.

Education has a special role to play in the context of the environmental crunch. As we said above: with a shift in attitudes, almost the same educational content in terms of the sciences and the arts could be deployed in a much less damaging way for the environment, by changing the way things are made or used as well as what we spend money on as consumers. Such a shift in attitudes holds the greatest promise for the deferment of Earth Overshoot Day, as well as the greatest chance of avoiding the possibility of the environmental crunch escalating from 'Big Stink' to 'Great Crunch'. Education has never been more important. For, make no mistake, the financial crunch will turn out to be a pale shadow of its 'twin' environmental crunch, and to prevent it or indeed to deal with its consequences the generations to come will need all their powers.

A tale of two credit crunches

So, will the financial credit crunch bankrupt the environment? It could, though it has to be said that humanity was doing a pretty good job of bankrupting the environment without the financial credit crunch.

The financial credit crunch is a game-changing event for economists. It will change consumer behaviour. Past consumer behaviour created the environmental credit crunch, which is critical for environmentalists. Environmentalists want consumers to change their behaviour. Unfortunately, not all of the changes inspired by the financial credit crunch are sufficient to push Earth Overshoot Day further out. As we said in the chapter on health (Chapter 8), for the consumer, the main impact of the financial credit crunch is that a fall in income or credit (or the fear thereof) can change consumption patterns (the mix of goods and services income is spent on) as well as the absolute amount spent in any given year. The consequences of these changing consumption patterns are diverse in their environmental implications.

The financial credit crunch thus does induce some changes that help to reverse the environmental credit crunch. It also creates an economic environment that will worsen the environmental credit crunch. Both financially and environmentally what we have is a lot of change, at the same time, which is not necessarily complementary.

The most fundamental conclusion we can draw from the twin credit crunches is the need for policymakers to recognize the consequences of *both* credit crunches, and to consider the *dual* credit consequences of their policy actions. There must be acceptance that economic relationships of the past cannot be replicated in a financially and environmentally constrained world.

This should not be read as a pessimistic prognosis. Humanity has demonstrated considerable capacity for change. Managing finite resources is by no means impossible – indeed, economics is all about managing finite resources (and infinite desires). Rather, this book is an attempt to demonstrate how the financial credit crunch has made the environmental credit crunch a more complex problem. We do not believe that the environmental credit crunch is an insoluble problem.

If policymakers have the wisdom to listen to both economists and environmental experts we should be able to keep the bailiffs away.

References

Ewing, B., Reed, A., Galli, A., Kitzes, J. and Wackernagel, M. (2010) *Calculation Methodology for the National Footprint Accounts*, Global Footprint Network, Oakland, CA

Kay, J. (2003) *The Truth About Markets*, Penguin Books, London and New York

Bibliography

Addison, P. and Crang, J. (2010) *Listening to Britain*, The Bodley Head, London

Allen, R. (2009) *The British Industrial Revolution in a Global Perspective*, Cambridge University Press, Cambridge

Bauman, H. and Tillman, A. (2004) *The Hitchhiker's Guide to LCA: An Orientation in Life Cycle Assessment Methodology and Application*, Studentlitteratur AB, Lund, Sweden

Beveridge, W. H. (1928) *British Food Control*, Oxford University Press, Oxford

Blumer, H. (1969) 'Fashion: From class differentiation to collective selection', *Sociological Quarterly*, vol 10, pp275–291

Bowden, S. and Turner, P. (1993) 'The demand for consumer durables in the United Kingdom in the interwar period', *The Journal of Economic History*, vol 53, pp244–258

Bowles, N. (2005) *Nixon's Business*, Texas A&M University Press, College Station, TX

Burkeman, O. (2010) 'This column will change your life: When reasoning goes out of the window', *Guardian Weekend Magazine*, 27 November

Chapagain, A. and Hoekstra, A. (2008) 'The global component of freshwater demand and supply', *Water International*, vol 33, no 1, pp19–32

Clark, D. (2004) *The Rough Guide to Ethical Shopping*, Rough Guides Ltd, London

De Vries, J. (2008) *The Industrious Revolution,* Cambridge University Press, New York

The Economist (2009) *Pocket World in Figures, 2010 Edition*, Profile Books Limited, London

Ewing, B., Reed, A., Galli, A., Kitzes, J. and Wackernagel, M. (2010) *Calculation Methodology for the National Footprint Accounts,* Global Footprint Network, Oakland, CA

Ford, W. (1882) *Dear Food*, Bradstreet, New York

Friedman, T. (1999) *The Lexus and the Olive Tree,* Farrar, Straus and Giroux, New York

Gautier, C. (2008) *Oil, Water and Climate*, Cambridge University Press, Cambridge

Georgescu-Roegen, N. (1971, 1999) *The Entropy Law and the Economic Process*, Harvard University Press, Cambridge, MA

Hardin, G. (1993) *Living Within Limits*, Oxford University Press, Oxford

Helm, D. and Hepburn, C. (eds) (2009) *The Economics and Politics of Climate Change*, Oxford University Press, Oxford

Hewitt, M. (2008) 'Earthships', *Green Building Bible*, vol 1, pp266–271

Hirsch, F. (1977) *Social Limits to Growth*, Routledge and Kegan Paul, London

Jackson, T. (2009) *Prosperity Without Growth*, Earthscan, London

Jenkinson, T. (ed) (1996) *Readings in Microeconomics*, Oxford University Press, Oxford

Jensen, M. C. and Meckling, W. H. (1994) 'The nature of man', *Journal of Applied Corporate Finance*, vol 7, no 2, pp4–19

Jevons, W. S. (1866) *The Coal Question: An Inquiry Concerning the Progress of the Nation, and the Probable Exhaustion of Our Coal-Mines*, Dodo Press, Gloucester

Kay, J. (2003) *The Truth About Markets*, Penguin Books, London and New York

Kerr, W. and Ryan, C. (2001) 'Eco-efficiency gains from remanufacturing: A case study of photocopier remanufacturing at Fuji Xerox Australia', *Journal of Cleaner Production*, vol 9, pp75–81

Keynes, J. (1936) *The General Theory of Employment, Interest and Money*, Macmillan Royal Economic Society edition (1973), London

Kolbert, E. (2006) *Field Notes from a Catastrophe*, Bloomsbury, New York

Lang, T., Barling, D. and Caraher, M. (2009) *Food Policy: Integrating Health, Environment and Society*, Oxford University Press, Oxford

Larsson, M. (2009) *Global Energy Transformation*, Macmillan, London

Lebergott, S. (1996) *Consumer Expenditures*, Princeton University Press

Leffall, L. and Kripke, M. (2010) *Reducing Environmental Cancer Risk*, US National Cancer Institute, Bethesda, MD

Light, A. (2007) *Mrs. Woolf and the Servants*, Penguin Books, London

Lovelock, J. (2006) *The Revenge of Gaia*, Allen Lane, London

Lyall, S. (2008) *The Anglo Files*, W. W. Norton and Company, London

MacKay, D. J. (2009) *Sustainable Energy – Without the Hot Air*, UIT, Cambridge

Magnus, G. (2009) *The Age of Aging*, John Wiley, London

Malthus, T. R. (1992) *An Essay on the Principle of Population*, edited by Donald Winch, Cambridge University Press, Cambridge

Mauser, W. (2008) *Water Resources: Efficient, Sustainable and Equitable Use* (English translation), Haus Publishing, London

Mayhew, H. (2010) *London Labour and the London Poor: A Selected Edition*, edited by Douglas-Fairhurst, R., Oxford University Press, Oxford

McClelland, P. and Tobin, P. (2010) *American Dream Dying*, Rowman and Littlefield, New York

McWilliams, J. E. (2009) *Just Food*, Little, Brown and Company, New York

Mokyr, J. (2009) *The Enlightened Economy*, Yale University Press, New Haven, CT

Norberg, J. (2009) *Financial Fiasco*, Cato Institute, Washington, DC

Olney, M. (1991) *Buy Now, Pay Later*, University of North Carolina Press, Chapel Hill, NC

Packard, V. (1961) *The Hidden Persuaders*, Penguin Books, London

Pensions Commission (2005) *A New Pensions Settlement for the Twenty-First Century: The Second Report of the Pensions Commission*, The Stationery Office, London

Peschardt, M. (1999) 'Australia "drowning in salt"', BBC News, 29 June, http://news.bbc.co.uk/1/hi/sci/tech/380907.stm

Porritt, J. (2005) *Capitalism as if the World Matters*, Earthscan, London

Pullen, T. (2008) *Simply Sustainable Homes*, Ovolo Publishing, Huntingdon

Rance, C. (1882) *The Water Supply of England and Wales*, Stanford, London

Ridley, M. (2010) *The Rational Optimist*, Fourth Estate, London

Russel, A. (1864) *The Salmon*, Edmonston and Douglas, Edinburgh

Sambrook, P. (1999) *The Country House Servant*, Sutton Publishing Limited, Thrupp, Stroud

Shiklomanov, I. and Rodda, J. (eds) (2003) *World Water Resources at the Beginning of the 21st Century*, Cambridge University Press, Cambridge

Smith, D. (2008) *The State of the World Atlas*, 8th edition, Earthscan, London

Smith, S. and Searle, B. (eds) (2010) *The Economics of Housing*, Wiley-Blackwell, Oxford

Solomon, S. (2010) *Water*, HarperCollins, New York

Strasser, S., McGovern, C. and Judt, M. (1998) *Getting and Spending*, Cambridge University Press, Cambridge

Sunstein, C. (2007) *Worst-case Scenarios*, Harvard University Press, Cambridge, MA

Sykes, F. (1944) *This Farming Business*, Faber & Faber, London

Touffut, J. (ed) (2009) *Changing Climate, Changing Economy*, Edward Elgar Publishing Ltd, Cheltenham

Twiss, T. (1844) *Two Lectures on Machinery*, Irish University Press, Shannon

UK Valuation Office Agency (2010) *The Agricultural Land and Property Market* and *The Residential Building Land Market*, www.voa.gov.uk/publications/property_market_report/pmr-jan-2010/index.htm

Vickery, A. (1998) *The Gentleman's Daughter*, Yale University Press, New Haven, CT

Ward, J. (1974) *The Finance of Canal Building in Eighteenth-Century England*, Oxford University Press, Oxford

The Westminster Review (1833) 'Poor laws commission', London

Wiedmann, T., Minx, J., Barrett, J., Vanner, R. and Ekins, P. (2006) *Sustainable Consumption and Production – Development of an Evidence Base*, Stockholm Environment Institute and Policy Studies Institute, Stockholm and London

Wilkinson, R. (1973) *Poverty and Progress: An Ecological Model of Economic Development*, Methuen, London

Woudhuysen, J. and Kaplinsky, J. (2009) *Energise*, Beautiful Books Ltd, London

Wrigley, A. (1988) *Continuity Chance and Change: The Character of the Industrial Revolution in England*, Cambridge University Press, Cambridge

Index